BOEING B-17
B-29 & LANCASTER

Stewart Wilson

Original illustrations by Juanita Franzi

INTRODUCTION

The second book in our *Legends of the Air* series chronicles the three most important heavy bombers of World War II – the Boeing B-17 Fortress, Avro Lancaster and Boeing B-29 Superfortress. I have attempted to cover their development in detail and to present a broad overview of their operational histories.

This book was written during the course of 1995, a period which has seen the 50th anniversaries of VE and VJ Days. It is perhaps appropriate therefore that these three aircraft be the subject of a book written at this time due to their massive contribution to the war effort in general and the major part they played in bringing hostilities to a close. Without the bravery and initiative of the airmen who flew them, these bombers' careers would be meaningless, and it is to them that this book is dedicated.

The bomber which bore the brunt of the USAAF's campaign against Germany in World War II, the B-17 came to symbolise the allied offensive in the last three years of the European war.

Sent to Germany in vast numbers and during daylight, Boeing's most famous bomber was subject to heavy losses early on, a situation which improved with the introduction of long range escort fighters which could go with it all the way to Berlin and back to bases in the United Kingdom.

It could be said the B-17 suffered from performance and operational limitations imposed by its pre war basic design but production exceeded 12,000 in six major versions and the aircraft served the USAAF and others on all fronts during the war. The legendary status which surrounded the B-17 during the war and afterwards is testimony to its secure place in aviation history.

The best British bomber of World War II, the Lancaster grew out of the failure of its twin engined predecessor, the Manchester, and went on to develop into a versatile aircraft capable of carrying a bigger bomb load than any other aircraft of the era. The RAF used the Lancaster to bomb Germany by night, the combination of it and the USAAF's daylight campaign proving to be one of the key elements leading to eventual victory.

Apart from its exploits during the course of its normal operations, the Lancaster is perhaps best known for the many special raids in which it participated including the sinking of the *Tirpitz* and the Dams Raid as well as its ability to carry the ten ton 'Grand Slam' bomb, the heaviest bomb used during World War II.

Postwar, the Lancaster evolved into the Lincoln and the basic airframe found new life as a transport and engine test bed. Production exceeded 7,000 including several hundred built in Canada.

Although best remembered as the bomber which ended the war with Japan by dropping atomic weapons on Hiroshima and Nagasaki, the B-29's exploits as a conventional bomber should not be forgotten, particularly its extensive use against Japan's industrial infrastructure operating from bases in the Marianas Islands.

First flown in 1942, the B-29 was a technological marvel of its time and far in advance of any other bomber with features such as pressurisation and remotely controlled defensive guns. It was developed quickly and suffered early problems as a result, but then settled down to effectively perform the role for which it was designed.

Production of the B-29 was just under 4,000 to which should be added nearly 400 examples of its closely related postwar development, the B-50. The early 1950s saw the B-29 with another war to fight, this time in Korea, this campaign closing the combat history of one of aviation's most significant military aircraft.

I gratefully acknowledge the help of the following people during the preparation of this book: Tom Lebbesmeyer (Boeing Historical Archives), Neil Mackenzie, Philip J Birtles, Juanita Franzi, Mike Kerr, Eric Allen, Jim Thorn, Gerard Frawley and Maria Davey. I hope that my efforts and theirs provide something of interest for the reader.

Stewart Wilson
Buckingham 1995

Published by Aerospace Publications Pty Ltd (ACN: 001 570 458), PO Box 3105, Weston Creek, ACT 2611, publishers of monthly *Australian Aviation* magazine.
Production Manager: Maria Davey

ISBN 1 875671 17 X

CONTENTS

Front cover (top to bottom): B-17F Fortress 42-30073 'Ole Puss' of 96th BG/413th BS USAAF; B-29 Superfortress 42-65280 'Dina Might' of 504th BG USAAF; Lancaster B.I R5868/PO-S of 467 (RAAF) Squadron. *(not to scale)*

BOEING B-17
FLYING FORTRESS

Contrails, tight formation and broad daylight – the key elements of B-17 Flying Fortress operations over Germany in World War II.

BOEING B-17 FLYING FORTRESS

The bomber which bore the brunt of the USAAF's campaign against Germany in World War II, the B-17 Flying Fortress came to symbolise the allied offensive in the last three years of the European war.

Sent to Germany in vast numbers and during daylight, Boeing's most famous bomber was subject to heavy losses early on, a situation which improved with the introduction of long range escort fighters which could go with it all the way to Berlin and back from bases in the United Kingdom. The combination of USAAF Fortresses and B-24 Liberators bombing Germany by day and RAF Lancasters, Halifaxes and Stirlings attacking by night proved to be a decisive one in the battle to liberate Europe.

The B-17 suffered from performance and operational limitations imposed by its pre war basic design. However at the time of its introduction the B-17 was very advanced, combining performance, load carrying and defensive firepower capabilities far beyond those of any contemporary bomber.

Originally developed as a private venture as the Boeing Model 299, the Fortress's qualities were sufficiently recognised for the project to survive the early crash of the first prototype and be ordered into relatively limited production at a time when the USA was isolationist and military spending was not a high priority. America's involvement in World War II changed that with the nation's tremendous industrial might coming to the fore. Well over 12,000 B-17s were eventually built in six major versions at three factories.

The aircraft served the USAAF and other air arms on all fronts during the conflict. The legendary status which surrounded the B-17 during the war and afterwards – particularly due to its exploits over Europe – is testimony to its secure place in aviation history.

The B-17 story does not end with World War II. Postwar, it found work as a transport, trainer, air-sea rescue, early warning, anti submarine, drone, testbed, firebomber and survey aircraft, some civil registered examples earning their keep well into the 1970s.

Boeing Background

The hugely successful Boeing company we know today has a history stretching back to the years of World War I with the establishment of the first partnership which would develop into the aerospace giant which has for many years been at the forefront of the industry.

William E Boeing was born in Detroit, Michigan, on 1 October 1881, the son of a wealthy timber merchant. By the early 1900s Boeing was already established as a prominent businessman in his own right, based in Seattle, Washington, with interests in timber, real estate and boatbuilding. His fascination with aviation was kindled by his first flight, as a passenger in a Curtiss seaplane flying from Lake Washington in July 1914.

Boeing struck up a friendship with a US Navy officer, Commander G Conrad Westervelt, who was serving on detachment to a Seattle shipyard. The two men visited air shows and examined the business of aviation generally, deciding they could build better aircraft than those they had seen.

An informal partnership was established with the intention of building seaplanes, a workshop was erected on the shores of Lake Union, Seattle, and an example of one of the most modern types available – a Martin Model T two seat seaplane – was purchased. Design of the first B & W aircraft (from the initials of the two builders) was carried out from 1916 and engineering expertise was obtained free of charge from the University of Washington, reward for an endowment Boeing had previously made to the establishment.

A beautiful portrait of B-17F Fortress 42-5234 over Mount Rainer, Washington, a favourite photo spot for Boeing's lensmen over the years. (Boeing)

The arrangement between the two men had to be dissolved before the first design had been completed, however, as Westervelt was posted to the east coast of the USA.

Development of what became known as simply the B & W (Model 1) carried on, the aircraft taking to the air for the first time on 29 June 1916. Substantially based on the Martin T owned by Boeing, the B & W was a twin float two seat biplane powered by a 125hp (95kW) Hall-Scott A-5 inline engine. Its maximum takeoff weight was 2,800lb (1,270kg) and it was capable of a top speed of 75mph (120km/h). Named *Bluebird*, the aircraft performed satisfactorily although its unusual control system – a wheel topped control column in which the wheel operated the rudders and the ailerons and elevators were operated normally by the 'stick' portion of the system – was soon replaced by a more conventional system.

A second B & W (named *Mallard*) was flown in November 1916 and the type generated sufficient interest in

William E Boeing, co-founder of the company which bears his name and its first president. (Boeing)

the US Navy for it to encourage development of a primary trainer version. *Bluebird* and *Mallard* were sold to the New Zealand Government in 1918.

Meanwhile, in July 1916, Boeing had established a formal corporation under which his aviation activities could be performed, the Pacific Aero Products Company. This was changed to the Boeing Airplane Company on 26 April 1917.

The entry of America into World War I in April 1917 gave Boeing its first production orders. The company's first entirely in house design – the Type C series of seaplanes – was responsible, the US Navy ordering 50 Type C Model 5s for training. Despite a low degree of reliability from the aircraft's 100hp (75kW) Hall-Scott A-7A engine, the US Navy's Type Cs served until the end of the war when they were sold off as war surplus, a couple of examples surviving until the early 1930s.

The war also brought with it a contract to manufacture 50 Curtiss HS-2L flying boats. This and the Model 5 order represented substantial contracts for the time and necessitated a move to larger premises, the company establishing a new facility

A photograph illustrating one of the vital components of the Allied success in World War II, America's industrial might. This is Boeing's Seattle plant in 1943 with acres of B-17s in production. (Boeing)

The first Boeing, the B & W seaplane of 1916. (Boeing)

at the Heath Shipyard, south of Seattle on the Duwamish River. Neither this nor the original Lake Union factory featured an airstrip, although this was not a problem at the time as the company was building floatplanes. It was not until 1928 that Boeing owned an airfield, Boeing Field (the former King County Airport) near Seattle.

The end of World War I saw half of Boeing's Curtiss HS-2L order cancelled and like other manufacturers affected by postwar cutbacks, it had to diversify to survive. Boeing returned to his boatbuilding origins for a time to keep the company busy, as well as manufacturing furniture and propellers.

This did not mean that aviation ac-

tivities were forgotten, with various prototypes produced and contracts for the refurbishment of Army and Navy de Havilland DH.4s fulfilled. As part of this, Boeing developed a new steel tube frame fuselage for the DH.4. The company won several contracts by undercutting others, one of these resulting in Boeing's biggest order to that point, for 200 Thomas-Morse MB-3A fighters in 1921. It was this project which turned Boeing around as it inspired the development of a new in-house fighter design, the PW-9 (Army) and FB-1 (Navy).

This family of biplane fighters was developed as a private venture and without official backing the risk was great. But it eventually paid off. The

first XPW-9 flew in June 1923 and the family spawned numerous variants between then and the end of the 1920s. Production of all PW/FB versions and their developments amounted to over 150 aircraft – a modest total but one on which Boeing based its subsequently successful fighter designs.

These included the F4B/P-12 series of biplanes, first flown in 1928 and manufactured in large numbers (nearly 700 for the Army, Navy and export); and the famous P-26 'Peashooter' all metal, open cockpit monoplane fighter, 138 of which were built for the Army Air Corps (plus 11 for export to China and one for Spain) between 1932 and 1936.

The Boeing Model 40 mailplane was built in the second half of the 1920s initially to meet a US Post Office Specification. Both mail and passenger carrying versions were built and production of all models reached 81 examples including four by Boeing's Canadian subsidiary. (Boeing).

The Civil Side

Boeing began a period of considerable expansion in the late 1920s based on the success of its fighters and the 1927 decision by the US Post Office to transfer transcontinental mail services from the government to the private sector. The company had already designed a specialist mailplane – the Model 40 biplane – two years earlier but its effectiveness was hampered by its heavy 400hp (300kW) Liberty engine. Installation of a much more efficient Pratt & Whitney Wasp or Hornet radial on the Models

40A, B and C from 1927 helped Boeing win the contract for the San Francisco-Chicago Mail service.

A new company, Boeing Air Transport Inc was formed to operate the service, flying from the newly acquired Boeing Field at Seattle. Further expansion followed with the acquisition of Pacific Air Transport, the new combination called The Boeing System. Other new Boeing ventures of the late 1920s included purchase of the Hamilton Metalplane Company, the establishment of a Canadian subsidiary (Boeing Aircraft of Canada

Ltd) in conjunction with Vancouver's Hoffar-Breeching Shipyard and the formation of the company's own school of aeronautics.

Perhaps the biggest area of expansion was when Boeing joined forces with Chance Vought, Hamilton Aero and Pratt & Whitney under the 'umbrella' holding company United Aircraft and Transport Corporation (UATC). This group was later joined by Sikorsky, Stearman, three airlines and the Standard Steel Propeller Company, which would subsequently merge with Hamilton to form the

The US Navy's F4B (top) and the Army Air Corps' P-12 series of biplane fighters of the late 1920s and 1930s. Combined production of these successful aircraft topped 700 including many for export. Illustrated are an FB4-1 and P-12B. (Boeing)

The Model 200 Monomail of 1930 incorporated some modern features such as retractable undercarriage, a semi monocoque fuselage and cantilever wing. (Boeing)

famous Hamilton Standard propeller firm. Even though the UATC members operated independently, new US antitrust laws soon came into play, stipulating that aircraft manufacturing and airline interests must be separated. This resulted in the July 1934 split of UATC into three parts – the Boeing Airplane Company (incorporating Stearman as a subsidiary), United Aircraft Corp (the other manufacturers) and United Air Lines Transport Corp.

In the meantime, production of the Model 40 reached 81 units (including four in Canada) and these were followed by a single example of the Model 200 Monomail, an innovative all metal cantilever low wing monoplane design with retractable undercarriage, monocoque fuselage construction and advanced aerodynamics. The sole Model 200 first flew in May 1930 and production was not under-

taken mainly due to the lack of a suitable variable pitch propeller which could fully exploit the aircraft's features. The propeller's pitch had to be set on the ground before takeoff and the choice had to be made then as to what performance characteristic was required – high payload and low performance or high performance and no payload!

A slightly lengthened version was also produced, the Model 221A Monomail with a claustrophobic cabin for six immediately behind the single 575hp (430kW) P&W Hornet engine. It was flown in August 1930 and subsequently modified to Model 221A standards with a further stretch of the forward fuselage and accommodation for two more passengers. The original Model 200 was also upgraded to this standard.

Many of the Monomail's advanced features would find application in

subsequent Boeing models including the B-9 bomber of 1931 and the Model 247 ten passenger twin engined airliner. First flown in February 1933, the 247 was state-of-the-art for the time: a low wing monoplane of all metal stressed skin construction. It was some 60mph (96km/h) faster than the trimotors it was intended to replace, cruising at around 180mph (290km/h) on the power of two 550hp (410kW) P&W Wasp radial engines.

Despite this and a good number of early orders – 60 from UATC 'family' member United Air Lines – the 247 was built in only small numbers, a total of 75 leaving the factory. The problem was the even more innovative Douglas DC-2 and then DC-3 series with their larger seating capacities and better economics. The DC-3 and its military derivatives went on to become the most produced transport aircraft in the history of

Boeing's first successful monoplane fighter, the P-26 'Peashooter' of the early 1930s. Despite being a monoplane it retained many old features including externally braced wings, an open cockpit and fixed undercarriage. This is a P-26B. (Boeing)

Towards the modern airliner (and bomber) – the Boeing 247. Despite its advanced features the 247 was overshadowed by the brilliance of the Douglas DC-2 and DC-3.

aviation. That the basic concept of the 247 was correct is not in doubt, but it was completely overshadowed by the great Douglas designs.

The performance of the Model 247 entered in the 1934 MacRobertson race between England and Australia probably sums the situation up nicely. Flown by Roscoe Turner and Clyde Pangborn, the Boeing did well, finishing third outright and second in the transport category. The overall winner was the purpose built de Havilland Comet racer, but second outright and first in the transport section was a Douglas DC-2.

The American Bomber

Despite some of the results achieved by bombing in World War I, the idea of strategic bombing was still very much a radical one during the 1920s and into the 1930s. This method of warfare had its supporters, but for many nations 'air forces' were still something of a novelty and probably dispensable in the time of 'peace' after 1918.

The United States was notably anti air force, or more accurately pro army and navy. Indeed, the US Army Air Corps (and Army Air Force from June 1941) did not achieve full independence as the US Air Force until 1947. As for strategic bombing, in the USA the power and influence of the Navy reigned supreme and any efforts to establish such a force were vigorously fought by the senior service, which considered it had some kind of divine right to

take care of America's offshore military needs.

The greatest advocate of the bomber in the USA was Brigadier General William ('Billy') Mitchell, a man who had seen the potential of aerial bombardment whilst an air commander of the American Expeditionary Force in France in 1917-18. To Mitchell, the establishment of a strategic bombing force was ideal for America's situation, geographically isolated as it was from the rest of the world. Mitchell's theories were based

William ('Billy') Mitchell, the USA's greatest bomber advocate.

on the idea that a highly mobile force of bombers would be far more effective responding to any attack than slow and unwieldy battleships or the movement of a large body of men. In other words, bombers could be the nation's first line of defence and could be used successfully against ships, something which raised the ire of the US Navy considerably.

As for the general theory of strategic bombing, Mitchell wrote: "To gain a lasting victory in war, the hostile nation's power to make war must be destroyed – this means the factories, the food producers, even the farms, the fuel and oil supplies, and the places where people live and carry on their daily lives ..."

Mitchell pressed his points so vociferously he was court martialled in 1925 and forced to resign his commission, the straw that broke the camel's back being a vicious and public attack on his superiors. This resulted from the September 1925 loss of a Navy airship along with 14 lives, Mitchell accusing the US War and Navy Departments of incompetence, criminal negligence and "treasonable administration of the national defence". These points were not readily accepted!

But Mitchell didn't stop with mere words. As assistant chief of the Air Corps he was in a position to demonstrate his theories from time to time. In 1921 he was given Congressional approval to perform a series of tests involving the bombing of former German warships which were anchored off the Virginia Capes. In July 1921, Martin MB-2 bombers sank the cruiser *Frankfurt* with 600lb (270kg) bombs; a destroyer was attacked by SE.5A fighters carrying 25lb (11kg) bombs and finished off by 300lb (136kg) bombs; and the battleship *Ostfriesland* was sunk by 1,000 and 2,000lb (454 and 907kg) bombs.

The finale to these exercises occurred in September 1921 when MB-2s dropped phosphorous, tear gas and incendiary bombs on the obsolete US battleship *Alabama* and finished her off with a 2,000 pounder. The object of the exercise was to "demonstrate how chemical missiles could make the decks and below decks of any war vessel a floating hell". Further attacks were made against obsolete battleships in 1923, the US Navy remaining officially unimpressed

Although the court martial saw him officially quietened, Mitchell's views had begun to win favour with many and by the time of his death in 1936 he was able to see that the basis of a US strategic bombing force had been established. Later, he was regarded as a national hero rather than villain and had the North American B-25 Mitchell medium bomber named after him.

Showing considerably more foresight than the politicians, the aircraft manufacturers were among those who could see the value in Mitchell's ideas and many of them theoretically applied the period's rapid advances in aircraft design to bombers, assuring that when the time came for the building of these aircraft, there would be some infrastructure in place.

The Army Air Corps achieved some degree of independence by being made responsible for all land based air defence of the USA and her overseas territories and aircraft up to a speed of 200mph (322km/h) and maximum bomb load of 2,000lb (907kg) were permitted. This action was approved by the US Navy which failed to see the full implications. The limitations and the Air Corps' charter of operations were broad enough to allow the development of a strategic bomber force, which if successful, would usurp part of the Navy's jealously guarded role almost by sleight-of-hand and under its very nose!

Another restriction was applied to American land based aircraft in 1938, again largely based on the Navy's concerns about traditional roles being taken from it. This time, aircraft were restricted to operations within 100 miles (161km) of the American coast, a totally impractical situation which brought the whole issue to a head. The tide had been irrevocably turning through the 1930s in favour of the bomber as time and time again their range capabilities were demonstrated with long flights.

As the available technology grew, so did aircraft speed, range, load carrying ability and reliability. By the late 1930s the Boeing B-17 Flying Fortress was on the scene, the apparent inevitability of war in Europe was causing everyone to rethink their defence positions, and the era which would either prove or disprove the strategic bomber advocates' theories was about to start.

Boeing's Bombers

Boeing's first attempt at a modern bomber was the Model 214/215, developed as a private venture in 1930. The different model numbers reflected different powerplants, the 214 having two 600hp (450kW) Curtiss GIV-1510C liquid cooled vee-type engines and the 215 a pair of 575hp (430kW) Pratt & Whitney R-1860 Hornet air cooled radials. The airframe represented a stepping stone between the old and the new. An all metal cantilever monoplane, the aircraft featured semi retractable undercarriage and for the first time on an American warplane, control surface servo tabs.

Some features from an earlier era were also included, the main one being open cockpits for the five crew members and an externally carried bomb load of up to 2,200lb (998kg). Defensive armament comprised two pairs of ring mounted 0.50in machine guns, one pair in the nose and another in a dorsal position. The aircraft was designated the B-9 by the USAAC and despite its mix of features exhibited high performance which matched or exceeded the biplane fighters of the day including a top speed of 186mph (299km/h). Its range was 1,150 miles (1,850km).

The Pratt & Whitney powered Model 215 (as the YB-9) was the first to fly in April 1931, followed in November 1931 by the Model 214 as the Y1B-9A. This aircraft was subsequently converted to Pratt & Whitney power. Only five more aircraft were built as Y1B-9As (Model 246) and used for evaluation. Delivered between July 1932 and March 1933, these were also powered by Pratt & Whitney Hornet engines but differed from the earlier aircraft by featuring a reprofiled rudder, reduced defensive armament, structural and equipment modifications and a maximum bomb load of 2,400lb (1,088kg).

The B-9 failed to achieve series production due to the appearance of the Martin B-10. Even more advanced than the Boeing aircraft, the B-10 had enclosed accommodation for its crew, variable pitch propellers and an internal bomb bay. The twin

Boeing's first attempt at an effective bomber, the B-9. Only seven were built between 1931 and 1933 but the aircraft was another step in the evolutionary process which would result in the B-17 Fortress. Illustrated is the first of the series to fly, the Model 215 YB-9. (Boeing)

engined B-10 became the standard Army Air Corps bomber of the 1930s.

The next Boeing bomber project to get underway was the enormous Model 294/XB-15, development of which began in 1934, the same time the smaller Model 299 (B-17 Flying Fortress) programme started. The two bombers were developed in parallel and entirely separately, the sole XB-15 not flying for the first time until October 1937, more than two years after the Model 299/B-17. The XB-15 is described in more detail in the B-29 Superfortress section of this book.

Forging the Fortress

The chain of events which led to the development of the B-17 Flying Fortress began in the first half of 1934 when the Army Air Corps began to solicit tenders for a new bomber to replace the Martin B-10, calling for a multi engined bomber capable of carrying one ton (1,000kg) of bombs over a distance of 2,000 miles (3,220km) at 200mph (321km/h). The prototype had to be flying within 12 months.

At this point Boeing had just been awarded a contract to build the XB-15 prototype and the company immediately decided to take a serious look at contesting the new competition. This was in view of a potential lack of other work with the P-26 fighter programme running down, the XB-15 heading towards being a one off and the 247 airliner already beginning to succumb to the Douglas DC-2 in terms of commercial sales despite a promising start. It would soon have to face the DC-3. Work was therefore begun on the Boeing Model 299 bomber, at this point very much as a private venture.

The Air Corps issued a more formal specification (Circular 35-26) for the new bomber in August 1934. Among its requirements: a maximum speed of 250mph (402km/h) at 10,000 feet; a 170-220mph (273-354km/h) cruising speed at the same height; an endurance of six-ten hours; a service ceiling of 20,000-25,000 feet (6,100-7,600m) and an engine out ceiling of at least 7,000 feet (2,133m). The exact number of engines was not specified, the requirement simply calling for a "multi engined four to six place land type airplane".

Although the Boeing board did not formally approve development of the Model 299 until September 1934, component manufacturing had begun the previous month. Drawing on its previous experience with all metal monoplanes and the work it was doing on the XB-15 project, Boeing's design team under the direction of Claire Egtvedt, project engineer E Gifford Emery and his assistant, Ed Wells (who was soon promoted to project engineer at the ripe old age of 24), came up with a low wing, all metal monoplane powered by four supercharged 750hp (560kW) Pratt & Whitney R-1690E Hornet radial engines. All the modern design features were there: retractable undercarriage (including the tailwheel), fully enclosed crew accommodation with side by side seating for the pilots, and Hamilton Standard constant-speed propellers.

Other design features included a wing structure with tubular truss spars, four welded aluminium fuel tanks with a total capacity of 1,700US gal (6,435 l) installed between the spars, a monocoque fuselage, wing flaps, fully cowled engines, tab controls and control surface gust locks which could be applied from the cockpit. The latter feature would later provoke fatal consequences early in the first Model 299's flying career.

By the standards of the time, the Model 299 as proposed was a large aeroplane, spanning nearly 104 feet (31.7m) and with a length of just under 62 feet (18.9m). The aircraft's 1,420sq ft (131.9m²) wing remained unchanged throughout its life despite

The first page of the US Army Air Corps specification calling for a new bomber to replace the Martin B-10. This prompted Boeing to go ahead with the company funded Model 299 four engined bomber. (Boeing)

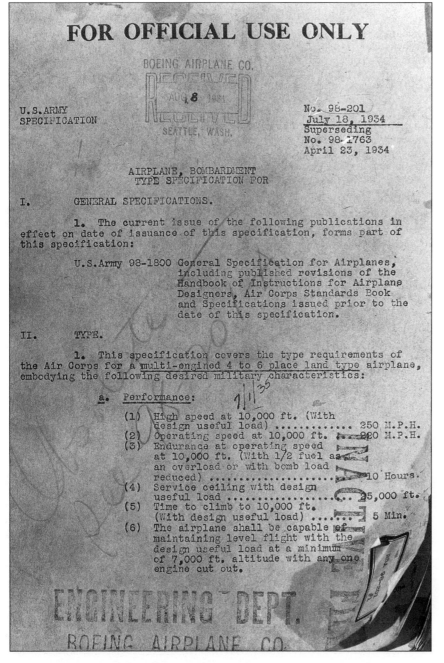

The Boeing Model 299, prototype for the B-17 series which would follow. Although it only had a brief life, the aircraft made an indelible impression on those who evaluated it. (Boeing)

just about everything else being redesigned or modified as time and operational demands progressed. Design gross weight was originally intended to be 28,000 to 30,000lb (12,700 to 13,600kg) with provision for an overload weight of around 35,000lb (15,900kg). In the event, the first Model 299 recorded a maximum weight of 43,000lb (19,500kg) and came in at 21,600lb (9,800kg) empty.

Defensive armament comprised five 0.3in machine guns, single units mounted in the nose (behind a Plexiglass nosepiece) plus dorsal, ventral and waist (one each side) positions in blisters. A maximum bomb load of 4,000lb (1,814kg) could be carried in a bomb bay behind the main crew compartment with bombs stacked vertically in two rows. This narrow bomb bay with a small opening area would later prove restrictive as it was too small to carry most of the larger bombs used by the Allies during World War II.

The Boeing board's September 1934 decision to go ahead with the Model 299 brought with it great financial risk as the cost of building a prototype was to be borne entirely by the company. Much of the initial $US275,000 allocated to the project had to be borrowed, while the final bill to develop the aircraft was over $US660,000, an enormous sum in the mid 1930s. This wasn't the last time Boeing would successfully 'bet the company' on a new aircraft either!

A Class Of Its Own

Considering the size and myriad of advanced features incorporated in the Model 299, it was a remarkable achievement for Boeing when the aircraft – carrying the civil experimental registration X-13372 – was rolled out

at Boeing field on 17 July 1935, just 11 months after full scale work on the project had begun. Taxying trials followed, during which a few minor problems were sorted out including an annoying tail wheel shimmy which took considerable effort to satisfactorily solve.

The rollout of this, the world's first all metal, four engined monoplane bomber, was a public affair in front of some members of the press including a Mr Richard Williams of the *Seattle Daily Times* newspaper. Such was the impression the new aircraft made on him, Mr Williams was moved to write about the "15 ton flying fortress" he had just seen. The words 'flying fortress' struck a chord within the company and that was the name bestowed on the new creation. 'Flying Fortress' was subsequently adopted by Boeing as a registered name, although in US and overseas military service the bomber was officially dubbed simply 'Fortress'.

First flight of the Model 299 took place from Boeing Field on the early

morning of 28 July 1935 with Boeing's chief test pilot Leslie Tower in command, Louis Wait as copilot and Pratt & Whitney's Henry Igo acting as flight engineer. There were no major problems and the crew reported enthusiastically on the aircraft's behaviour.

A limited flight test programme followed, after which the 299 would be evaluated in a 'fly off' competition against its two rivals for the Air Corps order in August. The other competitors were both smaller and less capable twin engined aircraft, a modified Martin B-10 and the Douglas DB-1, a hybrid combining the wing, powerplants and tail unit of the DC-2 airliner with a new fuselage. On paper at least, the 299 looked far superior in every respect to its rivals, its main drawback being its price, which at about $US197,000 per aircraft was twice that of the other two bombers.

Boeing's initial flight testing lasted just three weeks and comprised only seven flights totalling 14 hours and 15 minutes in the air. Limited mainly by a lack of time before Air Corps

A ground view of the Boeing 299. It first flew on 28 July 1935 but was destroyed in a crash just three months later.

evaluation was due to begin (and possibly a shortage of funds) the tests nevertheless revealed the 299's performance which included a top speed of 236mph (380km/h), a service ceiling of 24,600 feet (7,500m) and a normal range of 2,400 miles (3,860km). Only the maximum speed was below estimates.

Evaluation of the three contenders for the Air Corps order took place at Wright Field, Dayton Ohio, from 20 August. A major coup was planned for the 299 even before the trials began, the aircraft undertaking the 2,100 miles (3,380km) flight to Dayton non stop from Seattle. This was achieved in the time of 9hr 3min, at an average speed of 232mph (373km/h). In one stroke, the Boeing team had stolen a march on its rivals by forcefully demonstrating the 299's range and speed performance, the latter better than any fighter in USAAC service at the time.

The flyoff quickly proved the Boeing bomber was vastly superior to the Martin and Douglas offerings in every area except price. The Air Corps could not help but be impressed and all seemed to be going swimmingly for the Boeing 299.

Disaster

All the confidence engendered by the obvious success of the aircraft came to a grinding and tragic halt on 30 October 1935. On that day, X-13372 took off from Wright Field on a routine test flight with Major Ployer Hill (chief of the Wright Field Flight Test Section), Leslie Tower and others on board. Hill was flying the aircraft, which pitched up, stalled and crashed just after takeoff. The forward section

The Douglas B-18, winner of the 1935 USAAC bomber contest mainly by default due to the Boeing 299's crash. Despite the setback, Boeing's bomber had sufficiently impressed to ensure an evaluation batch order.

of the bomber was destroyed by fire, Hill was killed immediately and Tower later died of injuries received.

The cause was found to be the elevator gust locks which were operated from the cockpit. Hill had failed to remove them before takeoff, this serious enough situation being exacerbated by the reverse control action of the tabs moving on the locked surfaces.

The loss of two lives was the worst part of the tragedy, but for a time it looked as if the accident could also prove to be fatal for Boeing, remembering that it had everything tied up in the success of the Model 299. With only one prototype, Boeing had no other aircraft with which to complete the evaluation trials. As a result, the 299 was disqualified from the competition on that technicality at a time when the company was just about assured of receiving an order for 65

aircraft. The Air Corps was therefore more or less forced to order the next best contender, the vastly inferior Douglas DB-1 (now modified to incorporate DC-3 components), an initial contract for 133 being placed under the designation B-18.

Despite this serious setback, all was not lost for the Boeing 299 project. The aircraft's ground breaking capabilities had made a lasting impression on the evaluation team which could see its potential. Accordingly, in January 1936, a $US3.824m contract was awarded to the Seattle company for 13 aircraft (plus a static test airframe) for service evaluation under the designation YB-17.

This contract saved Boeing's financial bacon, as the prototype had cost it about $US500,000, even after insurance on the loss had been paid out. It was, as they say, a very close run thing.

Boeing Plant Two at Seattle, photographed in about 1937 with a Y1B-17 Fortress parked in front. (Boeing)

B-17F 42-5234 (top), the Fortress variant featured in the detailed cutaway drawings which follow over the next few pages. The first (bottom) illustrates the aircraft's nose section. (Boeing)

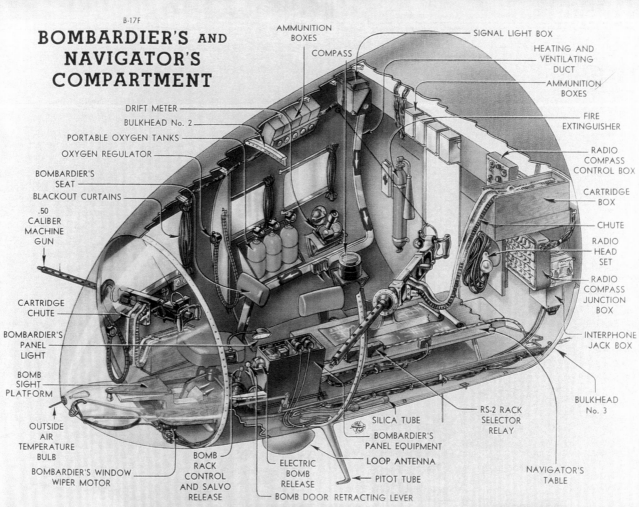

B-17F

BOMBARDIER'S AND NAVIGATOR'S COMPARTMENT

AMMUNITION BOXES

COMPASS

SIGNAL LIGHT BOX

HEATING AND VENTILATING DUCT

AMMUNITION BOXES

FIRE EXTINGUISHER

RADIO COMPASS CONTROL BOX

CARTRIDGE BOX

CHUTE

RADIO HEAD SET

RADIO COMPASS JUNCTION BOX

INTERPHONE JACK BOX

BULKHEAD No. 3

NAVIGATOR'S TABLE

RS-2 RACK SELECTOR RELAY

BOMBARDIER'S PANEL EQUIPMENT

LOOP ANTENNA

PITOT TUBE

SILICA TUBE

BOMB DOOR RETRACTING LEVER

ELECTRIC BOMB RELEASE

BOMB RACK CONTROL AND SALVO RELEASE

BOMBARDIER'S WINDOW WIPER MOTOR

OUTSIDE AIR TEMPERATURE BULB

BOMB SIGHT PLATFORM

BOMBARDIER'S PANEL LIGHT

CARTRIDGE CHUTE

.50 CALIBER MACHINE GUN

BLACKOUT CURTAINS

BOMBARDIER'S SEAT

OXYGEN REGULATOR

PORTABLE OXYGEN TANKS

BULKHEAD No. 2

DRIFT METER

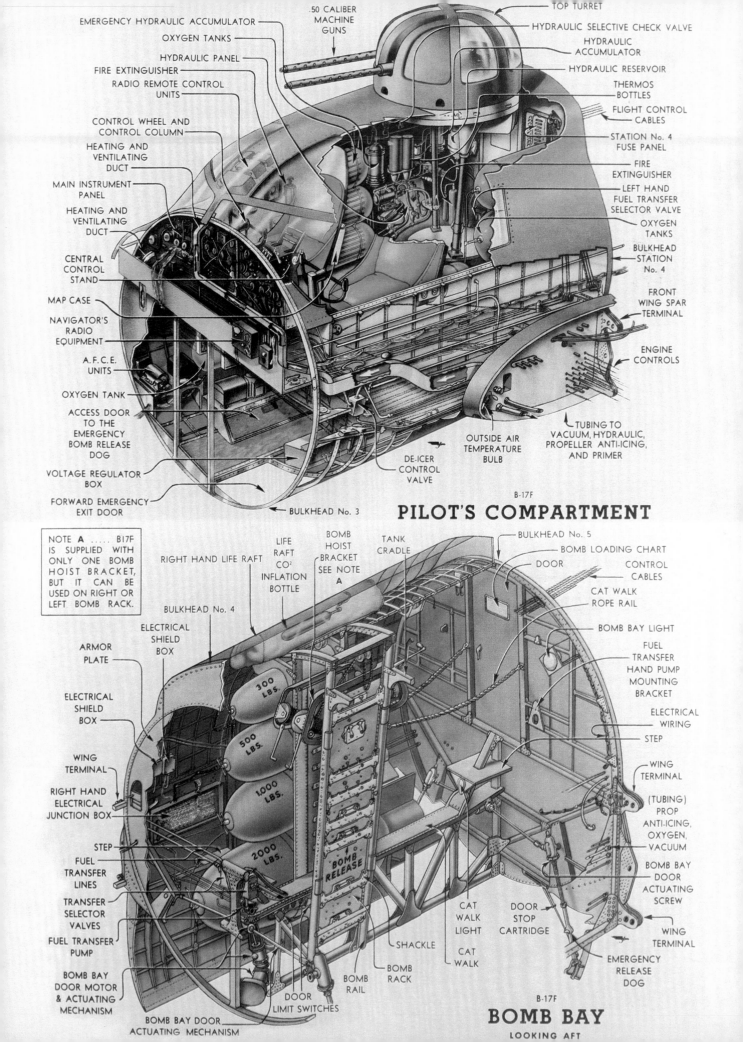

PILOT'S COMPARTMENT

.50 CALIBER MACHINE GUNS
TOP TURRET
HYDRAULIC SELECTIVE CHECK VALVE
HYDRAULIC ACCUMULATOR
HYDRAULIC RESERVOIR
THERMOS BOTTLES
FLIGHT CONTROL CABLES
STATION No. 4 FUSE PANEL
FIRE EXTINGUISHER
LEFT HAND FUEL TRANSFER SELECTOR VALVE
OXYGEN TANKS
BULKHEAD STATION No. 4
FRONT WING SPAR TERMINAL
ENGINE CONTROLS
TUBING TO VACUUM, HYDRAULIC, PROPELLER ANTI-ICING, AND PRIMER
OUTSIDE AIR TEMPERATURE BULB
DE-ICER CONTROL VALVE

EMERGENCY HYDRAULIC ACCUMULATOR
OXYGEN TANKS
HYDRAULIC PANEL
FIRE EXTINGUISHER
RADIO REMOTE CONTROL UNITS
CONTROL WHEEL AND CONTROL COLUMN
HEATING AND VENTILATING DUCT
MAIN INSTRUMENT PANEL
HEATING AND VENTILATING DUCT
CENTRAL CONTROL STAND
MAP CASE
NAVIGATOR'S RADIO EQUIPMENT
A.F.C.E. UNITS
OXYGEN TANK
ACCESS DOOR TO THE EMERGENCY BOMB RELEASE DOG
VOLTAGE REGULATOR BOX
FORWARD EMERGENCY EXIT DOOR
BULKHEAD No. 3

B-17F

PILOT'S COMPARTMENT

BOMB BAY

NOTE A .... B17F IS SUPPLIED WITH ONLY ONE BOMB HOIST BRACKET, BUT IT CAN BE USED ON RIGHT OR LEFT BOMB RACK.

RIGHT HAND LIFE RAFT
LIFE RAFT CO_2 INFLATION BOTTLE
BOMB HOIST BRACKET SEE NOTE A
TANK CRADLE
BULKHEAD No. 5
BOMB LOADING CHART
DOOR
CONTROL CABLES
CAT WALK ROPE RAIL
BOMB BAY LIGHT
FUEL TRANSFER HAND PUMP MOUNTING BRACKET
ELECTRICAL WIRING
STEP
WING TERMINAL
(TUBING) PROP ANTI-ICING, OXYGEN, VACUUM
BOMB BAY DOOR ACTUATING SCREW
WING TERMINAL
EMERGENCY RELEASE DOG

BULKHEAD No. 4
ELECTRICAL SHIELD BOX
ARMOR PLATE
ELECTRICAL SHIELD BOX
WING TERMINAL
RIGHT HAND ELECTRICAL JUNCTION BOX
STEP
FUEL TRANSFER LINES
TRANSFER SELECTOR VALVES
FUEL TRANSFER PUMP
BOMB BAY DOOR MOTOR & ACTUATING MECHANISM
BOMB BAY DOOR ACTUATING MECHANISM
DOOR LIMIT SWITCHES
BOMB RAIL
BOMB RACK
SHACKLE
CAT WALK LIGHT
CAT WALK
DOOR STOP CARTRIDGE

300 LBS.
500 LBS.
1,000 LBS.
2,000 LBS.
BOMB RELEASE

B-17F

BOMB BAY

LOOKING AFT

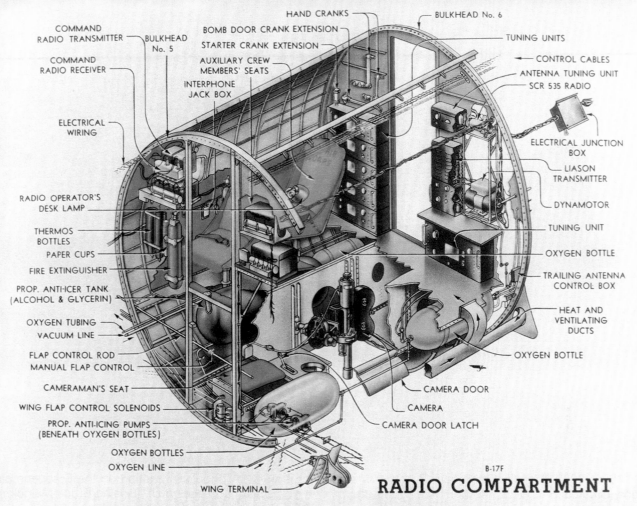

COMMAND RADIO TRANSMITTER
COMMAND RADIO RECEIVER
BULKHEAD No. 5
INTERPHONE JACK BOX
ELECTRICAL WIRING
HAND CRANKS
BOMB DOOR CRANK EXTENSION
STARTER CRANK EXTENSION
AUXILIARY CREW MEMBERS' SEATS
BULKHEAD No. 6
TUNING UNITS
CONTROL CABLES
ANTENNA TUNING UNIT
SCR 535 RADIO
ELECTRICAL JUNCTION BOX
LIASON TRANSMITTER
DYNAMOTOR
TUNING UNIT
OXYGEN BOTTLE
TRAILING ANTENNA CONTROL BOX
HEAT AND VENTILATING DUCTS
OXYGEN BOTTLE
CAMERA DOOR
CAMERA
CAMERA DOOR LATCH

RADIO OPERATOR'S DESK LAMP
THERMOS BOTTLES
PAPER CUPS
FIRE EXTINGUISHER
PROP. ANTI-ICER TANK (ALCOHOL & GLYCERIN)
OXYGEN TUBING
VACUUM LINE
FLAP CONTROL ROD
MANUAL FLAP CONTROL
CAMERAMAN'S SEAT
WING FLAP CONTROL SOLENOIDS
PROP. ANTI-ICING PUMPS (BENEATH OYXGEN BOTTLES)
OXYGEN BOTTLES
OXYGEN LINE
WING TERMINAL

B-17F
RADIO COMPARTMENT

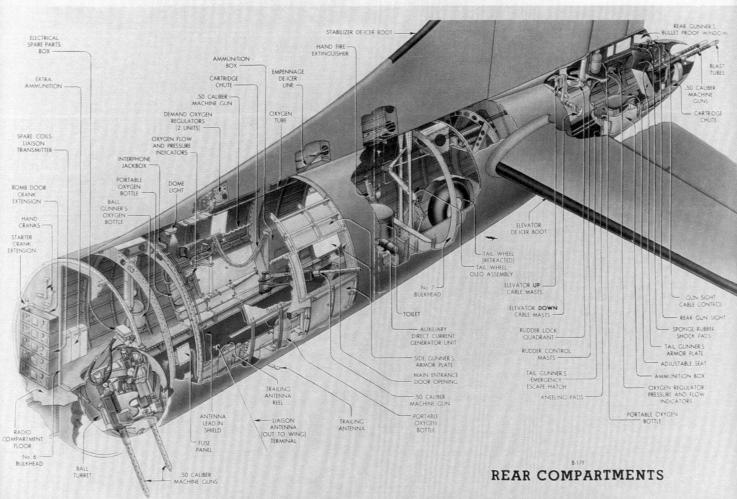

ELECTRICAL SPARE PARTS BOX
EXTRA AMMUNITION
SPARE COILS LIASION TRANSMITTER
BOMB DOOR CRANK EXTENSION
HAND CRANKS
STARTER CRANK EXTENSION
RADIO COMPARTMENT FLOOR
No. 6 BULKHEAD
BALL TURRET

AMMUNITION BOX
CARTRIDGE CHUTE
.50 CALIBER MACHINE GUN
DEMAND OXYGEN REGULATORS (2 UNITS)
OXYGEN FLOW AND PRESSURE INDICATORS
INTERPHONE JACKBOX
PORTABLE OXYGEN BOTTLE
BALL GUNNER'S OXYGEN BOTTLE
DOME LIGHT

STABILIZER DE-ICER BOOT
HAND FIRE EXTINGUISHER
EMPENNAGE DE-ICER LINE
OXYGEN TUBE

REAR GUNNER'S BULLET PROOF WINDOW
BLAST TUBES
.50 CALIBER MACHINE GUNS
CARTRIDGE CHUTE

ELEVATOR DE-ICER BOOT
TAIL WHEEL (RETRACTED)
TAIL WHEEL OLEO ASSEMBLY
ELEVATOR **UP** CABLE MASTS
ELEVATOR **DOWN** CABLE MASTS
RUDDER LOCK QUADRANT
RUDDER CONTROL MASTS
TAIL GUNNER'S EMERGENCY ESCAPE HATCH
KNEELING PADS

GUN SIGHT CABLE CONTROL
REAR GUN SIGHT
SPONGE RUBBER SHOCK PADS
TAIL GUNNER'S ARMOR PLATE
ADJUSTABLE SEAT
AMMUNITION BOX
OXYGEN REGULATOR PRESSURE AND FLOW INDICATORS
PORTABLE OXYGEN BOTTLE

No. 7 BULKHEAD
TOILET
AUXILIARY DIRECT CURRENT GENERATOR UNIT
SIDE GUNNER'S ARMOR PLATE
MAIN ENTRANCE DOOR OPENING
.50 CALIBER MACHINE GUN
PORTABLE OXYGEN BOTTLE

TRAILING ANTENNA REEL
ANTENNA LEAD-IN SHIELD
FUSE PANEL
LIASION ANTENNA (OUT TO WING) TERMINAL
TRAILING ANTENNA
.50 CALIBER MACHINE GUNS

B-17F
REAR COMPARTMENTS

B-17E 41-2600 (top) illustrated near Boeing's Seattle plant in 1942. A popular attraction on the European air show circuit is B-17G-VE 44-85784 'Sally B' (bottom), one of the last Fortresses built. (Boeing/Bruce Malcolm)

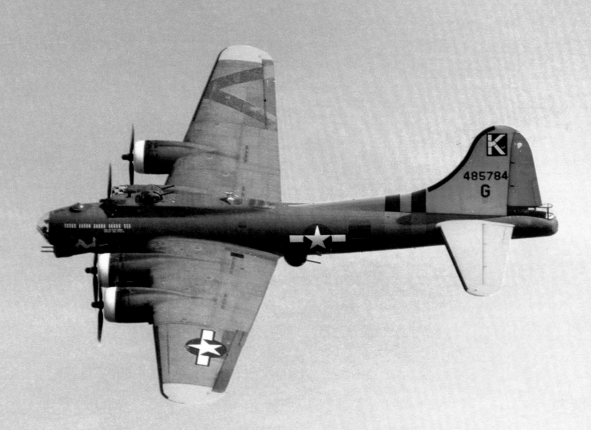

Postwar Fortresses: one of the two JB-17G engine test beds (top), this one originally converted to test the Curtiss-Wright Typhoon turboprop in the nose; and B-17G CF-HBP (bottom), the former 44-83814 in service with Kenting Aviation on survey duties between 1953 and 1971. (Eric Allen)

B-17G 'Sally B' (top) photographed at Duxford in 1975; and some interesting detail in this shot of the nose section of a Fortress (bottom), suitably 'battle damaged' for film work. (Eric Allen)

DEVELOPING THE FORTRESS

Y1B-17 FORTRESS

The US Army Air Corps' January 1936 decision to order 13 Fortresses for evaluation (plus one static test airframe) not only saved Boeing's financial bacon but also ensured a modern bomber would be in service and series production at the outbreak of World War II. By the time America entered the war at the end of 1941, the Army Air Force (as it was by then) had only nine Consolidated B-24 Liberators in service whereas the B-17 was already well established with about 150 built to that point and the first of the substantially redesigned and much improved 'new generation' Fortress – the B-17E – flying and entering large scale production.

The service evaluation Fortresses (Boeing Model 299B) were initially given the USAAC designation YB-17 but this was quickly changed to Y1B-17, the '1' in the designation indicating the aircraft was subject to special funding. The 13 aircraft were serialled 36-149 to 36-161 and the first of them was flown on 2 December 1936. The last was handed over in August the following year.

Externally they were similar to the Model 299 prototype, but under the skin they featured several substantial changes which further enhanced the capability demonstrated by the first aircraft. The major change was the replacement of the 299's 750hp (560kW) Pratt & Whitney Hornets with 1,000hp (750kW) Wright R-1820-39 Cyclone nine cylinder single row radial engines. The Cyclone remained the B-17's powerplant throughout its production life.

There were other more detailed changes to the aircraft: fuel capacity was increased, the oxygen system modified, the flaps were fabric instead of metal covered, the main undercarriage incorporated a single rather than dual oleo leg to facilitate wheel changes; defensive armament was improved and the maximum weight went up by 650lb (295kg) to 43,650lb (19,800kg). There were also equipment and internal arrangement changes and de-icing equipment could be fitted.

A vitally important piece of equipment which would be fundamental to the B-17's and other bombers' operational careers was the Norden bombsight, named after its inventor, Carl Norden. Developed in the 1920s, the sight comprised a system of gyros and mirrors which were designed to place bombs accurately on their targets.

The Norden allowed the bombardier to fly the aircraft during the bombing run, adjusting its course. The sight's crosshairs were locked onto the target and the speed and altitude of the bomber plus the bombs' ballistic data were entered into its computer system, all of this

Douglas built B-17G-30-DL, one of 2,395 'G' model Fortresses manufactured by the company at Long Beach. This aircraft well displays the B-17's broad wing and its dorsal, tail, chin, waist and cheek gun positions. The ventral ball turret is obscured. (Boeing)

Y1B-17s of the USAAC's 2nd Bombardment Group over New York in February 1938, on their way to South America on one of the several long distance flights which helped prove the worth of the new bomber. (Boeing)

combining to inform the bombardier when the bombs should be released, this action being performed automatically once the necessary information had been entered. Later development allowed the sight to be linked to the aircraft's autopilot during the bombing run, allowing course corrections to be automatically made. Not surprisingly, the Norden bomb sight remained highly secret throughout the war.

Despite its obvious capabilities, the Fortress had its share of early problems and at one stage the entire programme was once again threatened. This situation resulted from an accident involving the first Y1B-17

just five days after its maiden flight when it nosed over on landing due to the brakes seizing on. The pilot had caused the trouble by using the brakes excessively while taxying out and then retracting the undercarriage before the brakes had cooled. A Congressional Enquiry followed, but the B-17 programme was allowed to continue.

That same flight also highlighted some early problems with the Wright Cyclone engine. The landing which resulted in the Y1B-17 nosing over was actually an emergency one at Boeing Field following the overheating of two engines.

Proving The Concept

At this point the Fortress's future was still an unknown quantity. The politicians were unsure of it (and the role of bombers generally) and the US Navy was opposed to anything which might threaten its supreme position as Defender of All Things Offshore.

The Fortress had to prove itself. This it did through a series of highly publicised long distance flights and a public relations campaign carried out on its behalf by the Army Air Corps through General Headquarters Air Force, a body created in 1935 to develop a strategic bomber organisation within the USAAC. GHCAF was commanded by Brig Gen Frank Andrews, a staunch supporter of the long range bomber concept. The worsening situation in Europe in the late 1930s also helped sway some important opinion in favour of the bomber although stern opposition had to be overcome first.

The Y1B-17s were operated by the 2nd Bombardment Group at Langley Field, Virginia, with one aircraft held at Wright Field for experimental work. The 2nd BG quickly set about putting the capabilities of its new bomber before the public with long distance flights within the USA and overseas. Over a period of time these demonstrated the aircraft's reliability by flying 10,000 accident free hours in all weathers.

Goodwill missions were flown to Argentina and Brazil, while the Y1B-17s also broke the existing transcontinental USA records with ease, setting new east-west and west-east

A Y1B-17 Fortress of the 2nd BG.

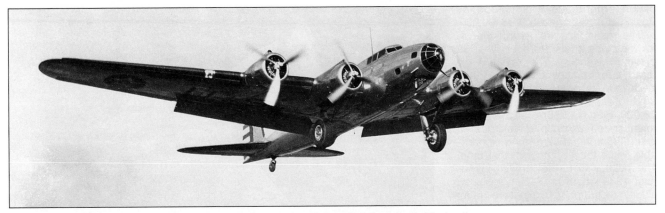

A Y1B-17 on short finals with its split flaps extended for landing.

times of 12hr 50min and 10hr 46min, respectively.

One particularly spectacular flight sparked off an unholy row between the Navy and Air Corps. This occurred in May 1938 when three of the 2nd BG's Y1B-17s intercepted the Italian liner *Rex* some 700 miles (1,125km) out to sea during a navigation exercise. Feeling threatened, the Navy demanded (and temporarily got) restrictions on Air Corps aircraft which meant they could not operate more than 100 miles (161km) out to sea. The furore also cost Andrews his job, although his stand on the subject would soon be vindicated.

As a sidelight, in the lead aircraft on the *Rex* interception flight was a certain Lt Curtis LeMay, a man who would shortly assume some importance in the world of strategic bombing.

One other incident set the scene for part of the B-17's future status as a legend of the air – its ability to absorb enormous amounts of punishment but still make it home. One of the Y1B-17s tested the aircraft's structural strength to the limit when it encountered extremely heavy turbulence near Langely. Flown by Lt William Bentley, the Fortress was thrown out of control and into a nine turn spin, these actions exerting forces on the airframe which were greater than it was supposed to stand. Bentley recovered control and

landed the aircraft, which was found to have suffered bent wings and other structural and cosmetic damage.

The Y1B-17 was repaired and flown again and as a result of this unrehearsed and unexpected natural structural test, the extra airframe which had been set aside for static structural testing was released for conversion into an airworthy Fortress as the Y1B-17A.

Y1B-17 FORTRESS

Powerplants: Four Wright R-1820-39 Cyclone nine cylinder supercharged single row radial engines rated at 1,000hp (750kW) for takeoff and 850hp (635kW) at 5,800ft; Hamilton Standard constant-speed and feathering three bladed propellers of 11ft 6in (3.50m) diameter; fuel capacity 1,730 USgal (6,549 l).
Dimensions: Wing span 103ft 9in (31.62m); length 68ft 4in (20.83m); height 15ft 4in (4.67m); wing area 1,420sq ft (131.9m²).
Weights: Empty 24,460lb (11,095kg); max loaded 43,650lb (19,800kg).
Armament: Normal bomb load 4,000lb (1,814kg), max bomb load 8,000lb (3,629kg); five 0.30in machine guns.
Performance: Max speed 208kt (384km/h) at 5,000ft, 222kt (412km/h) at 14,000ft; cruising speed 152kt (281km/h); time to 10,000ft 6.5min; service ceiling 30,600ft (9,327m); normal range 2,088nm (3,862km); max range 2,960nm (5,471km).

Y1B-17A FORTRESS

The release of the static test airframe for conversion to full flying status resulted in an additional aircraft becoming available to the evaluation fleet, the Model 299F Y1B-17A, serial number 37-369. This aircraft was probably responsible for finally convincing the US powers-that-be to the virtues of the long range bomber.

The Y1B-17A was first flown in April 1938 and it differed from its predecessors mainly in having R-1820-51 Cyclone engines fitted with Moss-General Electric turbosuperchargers. These would help transform the Fortress into a much more effective bomber due to its high altitude performance and form the basis for the versions which would shortly be seeing active service. The theory of high altitude precision bombing as favoured by many of the Air Corps' bomber advocates was much closer to being a practical reality thanks to the combination of turbosupercharged engines and the Norden bombsight.

Once again, there was a problem which had to be overcome before the full benefits of this new B-17 could be exploited. The problem stemmed from the fact that both the exhausts and turbosuperchargers were mounted above the engine nacelles – an Army requirement – flight testing immediately showing that this arrangement caused considerable aerodynamic problems, excessive turbulence due to the airflow over the wing being disturbed causing severe vibration and buffeting.

Boeing spent a substantial amount of time and money trying to solve the problem while leaving the exhaust and turbocharger fairings where the Army required them to be without success. The company finally had to re-route the exhaust so it and the turbochargers could be mounted below the nacelle, an expensive and time consuming exercise which required considerable reworking. The modifi-

A Y1B-17 of the 96th Bomber Squadron displays the early aircraft's distinctive nose design with gun mounting 'bubble' above the transparency and the bomb aimer's 'notch' on the lower surface.

cation programme was funded by Boeing at a cost of about $US100,000.

The result was well worth the effort, service trials from early 1939 proving the theory. The Y1B-17A offered substantially better performance than its predecessors including an increase in service ceiling to 38,000 feet (11,580m) and a maximum speed at optimum altitude of over 300mph (483km/h), a remarkable figure for a bomber of that time. The point was further proven in August 1939 when the Y1B-17A set an altitude record of 34,000 feet (10,360m) while carrying an 11,000lb (4,990kg) payload and also established a 621 miles (1,000km) closed circuit speed record of 259.4mph (417.4km/h).

From now on all B-17s (and all future piston engined US heavy bombers) would have turbosupercharged engines.

The sole Y1B-17A (37-369) with turbosupercharged engines. Built from an airframe originally intended for static structural testing, this aircraft was an important factor in convincing the US powers-that-be of the virtues of the long range bomber.

B-17B FORTRESS

The first production Fortress, the B-17B (Model 299E initially, then 299M) was basically similar to the Y1B-17A with its turbosupercharged Wright R-1820-51 engines but incorporating several detail changes. The most obvious was the revised and shortened nose shape which replaced the previous design (incorporating a bubble 'turret' for the nose machine gun) with a smoother plexiglass nose cone which included the bomb aimer's window and ball and socket gun mounts.

The new arrangement created more space for the bomb aimer and navigator. The cut out bomb aiming 'notch' in the lower part of the nose was replaced with a flat panel on which the Norden bomb sight was carried. An observation bubble for the aircraft commander was added to the top of the rear cockpit.

Other changes included a rudder of broader chord, larger flaps and a hydraulic instead of pneumatic braking system. Defensive armament remained as before and the normal crew complement was six. Maximum permissible weight further increased to 46,178lb (20,946kg).

An initial contract for 10 B-17Bs was placed in August 1937, before flight testing of the turbosupercharged Y1B-17A had begun. The performance advantages bestowed by this development and the decision to adopt it on production aircraft resulted in delays in getting the aircraft into production. The first example (38-211) flew on 27 June 1939 and was delivered to the Army Air Corps' 2nd Bombardment Group on 20 October.

Production of the B-17B Fortress eventually amounted to 39 aircraft, the last delivered in March 1940.

(below) The first production Fortress, the B-17B. Production amounted to 39 aircraft, the first aircraft flying in June 1939. Compare this shot with the B-17C photographed from a similar angle. (Boeing)

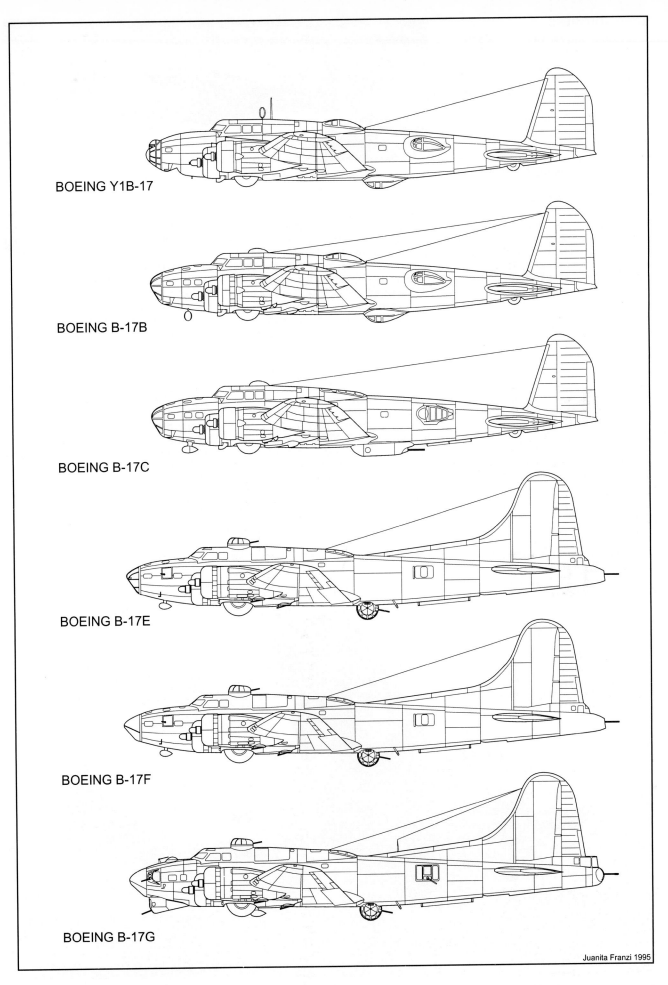

BOEING Y1B-17

BOEING B-17B

BOEING B-17C

BOEING B-17E

BOEING B-17F

BOEING B-17G

Juanita Franzi 1995

B-17C FORTRESS

Although the Air Corps was screaming out for more B-17s, orders continued to come in only modest quantities and at a fairly leisurely pace, the contract for 38 B-17Cs (serials 40-2042 to 40-2079) being placed only in September 1939, just over two weeks after war had been declared on Germany. Despite this, the neutral USA saw no reason yet for large scale rearmament and was able to divert 20 of the new B-17Cs to Britain for the Royal Air Force.

The first B-17C was flown on 21 July 1940 and deliveries were completed in November of the same year. The major change over the B-17B was the installation of R-1820-65 Cyclones with increased turbo boost and therefore more power under all circumstances, especially at altitude. Rated at 1,200hp (895kW) for takeoff, the R-1820-65 could maintain 1,000hp (745kW) at 14,200 feet in military rating giving the B-17C a maximum speed of 323mph (520km/h) at 25,000 feet, the fastest of all the Fortress variants. The RAF reputedly recorded a speed of 353mph (568km/h) with one of its B-17Cs, a truly remarkable achievement if true.

Other B-17C changes included removal of the waist gun blisters and their replacement with a flat panel opening which reduced drag and offered a better field of fire; the ventral gun blister was replaced with a

A B-17C undergoing testing against the familiar background of Mount Rainer. This model's revised nose shape is apparent. (Boeing)

smoother 'bathtub' design; the dorsal gun blister was replaced with a flush plexiglass fitting; 0.50in machine guns were mounted in all positions except the nose; twin 'fifties' subsequently appeared in the ventral and dorsal positions. Self sealing fuel tanks and some armour protection were later added to most aircraft, these and other modifications bringing them up to basically B-17D standards.

Fortresses For Britain

The USA's neutrality did not prevent President Roosevelt from sending arms to Britain to help in her fight against Germany and in April 1941 the first of 20 B-17Cs arrived in England. Given the RAF name Fortress I and serialled AN518-AN537, these were the first Fortresses to see action.

Flown by 90 Squadron from West Raynham, Norfolk, the Fortresses were used on several daylight raids

over Germany during 1941 before the survivors were put on anti-shipping and anti-submarine duties. The first raid performed by the aircraft was on Wilhelmshaven's submarine building docks on 8 July 1941. Berlin was attacked later in the month.

These operations are discussed in a later chapter, but they were generally unsuccessful for a number of reasons, including vulnerability to enemy fighters – particularly in frontal and tail attacks – something which made their 'Flying Fortress' name something of a mockery. Other shortcomings were also noted (some of which were due to the tactics employed) but the result was that it was recognised that the B-17 had a long way to go before it could be regarded as a first rate combat aircraft. Nevertheless, a great deal of knowledge was gathered which would be put to good use in later versions.

One of the 20 B-17Cs delivered to the Royal Air Force in 1941 under the name Fortress I. Note the serial number has been incorrectly painted on the aircraft – it should be AN528. (Boeing)

This shot of a B-17C can be compared with the similar angle on a B-17B. Visible external differences include the flat panel waist gun openings, the flush plexiglass dorsal gun blister and the 'bathtub' ventral gun fairing. (Boeing)

B-17C FORTRESS
Powerplants: *Four Wright R-1820-65 Cyclone nine cylinder, single row, turbosupercharged radial engines each rated at 1,200hp (895kW) for takeoff and 1,000hp (745kW) at 14,200ft; Hamilton Standard constant-speed and feathering three bladed propellers of 11ft 6in (3.50m) diameter; fuel capacity 1,730 US gal (6,549 l).*
Dimensions: *Wing span 103ft 9in (31.62m); length 67ft 11in (20.7m); height 15ft 4in (4.67m); wing area 1,420sq ft (131.9m²).*
Weights: *Empty 30,600lb (13,880kg); normal loaded 46,650lb (21,160kg); max loaded 49,650lb (22,521kg).*
Armament: *Normal bomb load 4,000lb (1,814kg); five or six 0.50in and one 0.30in machine guns.*
Performance: *Max speed 281kt (520km/h) at 25,000ft; max cruising speed 249kt 460km/h); time to 10,000ft 7.5min; service ceiling 37,000ft (11,277m); normal range 2,088nm (3,862km); max range 2,960nm (5,470km).*

B-17D FORTRESS

Forty-two further B-17s were ordered in April 1940, the changes incorporated in these Fortresses earning them the new designation B-17D. The aircraft were serialled 40-3059 to 40-3100 and were delivered between February and April 1941.

Externally, the major change between the B-17D and its predecessor was the incorporation of engine cowl flaps intended to improve engine cooling during prolonged climbing to altitude. The B-17D was originally intended to have Sperry dorsal and ventral powered turrets installed but as these were not ready in time the gun armament was increased to twin 0.50in guns in these positions. As part of their upgrade to 'D' model standards, most B-17Cs also received these installations.

Other improvements included self sealing fuel tanks and increased armour protection (also retrofitted to most Cs), the electrical system was

changed from 12 volt to 24 volt, an improved low pressure oxygen system was installed and the bomb racks were modified.

The B-17C/D was the first Fortress to see action in US colours. Twelve 5th Bombardment Group aircraft were on the ground at Hickham Field, Hawaii, on the morning of 7 December 1941 when Japan attacked Pearl Harbour and surrounding areas. Most of the Fortresses were destroyed while another 12 unarmed aircraft from the 7th BG arrived at Hickham in

the middle of the raid. One was lost and all the others were damaged. Nine hours later, Japan attacked the Philippines, destroying nine B-17s at Clark Field.

The survivors of the USAAF's B-17 force in the Pacific rallied to perform the Fortress's first offensive mission on 10 December when five aircraft attacked a Japanese convoy off the coast of Luzon. These were the first bombs dropped by American aircraft in World War II and although several hits were reported, no ships were sunk.

(above) The B-17D with cowl flaps introduced to improve engine cooling during prolonged climbing to altitude. (Boeing)

(below) The scene inside Boeing's Plant 2 at Seattle in March 1941 with B-17Ds under construction. Forty-two B-17Ds were built, bringing total production for the early generation Fortress to just 134 aircraft including the prototype and evaluation batch. (Boeing)

Mid production B-17E 41-2599 shows this model's substantially redesigned and lengthened rear fuselage, dorsal and ventral turrets and new tail surfaces.

B-17E FORTRESS

The first of the 'new generation' heavily redesigned Fortresses and the first to achieve quantity production, the B-17E incorporated numerous modifications – especially in the area of defensive armament – which would form the basis of the aircraft used in World War II.

Although these modifications on the whole reflected dealing with many of the problems encountered by the Royal Air Force in its operational use of early model B-17s during the second half of 1941, it would be wrong to suggest this influenced the B-17E's design, as by the time the RAF aircraft entered combat the design work had already been done. Rather, Boeing looked at the B-17C/D, identified

shortcomings and sought to overcome them. One area of British influence was the incorporation of powered gun turrets in three positions. It will be recalled that the B-17D was originally intended to have dorsal and ventral powered turrets installed but they were not sufficiently developed in time.

Developed under the Boeing Model 2990 designation (which would also cover subsequent Fortress models), the B-17E's fundamental change was a completely redesigned rear fuselage of greater width, length and depth, new vertical tail surfaces incorporating a large fin fillet and horizontal surfaces of increased span. The B-17E's overall length was 5ft 11in (1.80m) greater than before.

Defensive armament now comprised a single 0.30in machine gun mounted in the plexiglass nose, single 0.50in machine guns in the two waist positions (now firing through rectangular apertures), twin manually operated 'fifties' in the tail with the gunner housed in a compartment below the rudder, a powered Sperry dorsal turret mounted just behind the cockpit with a pair of 0.50in guns and a similarly armed Bendix remotely controlled powered ventral turret in early aircraft replacing the previous 'bathtub' arrangement.

This weapon was aimed and fired by a gunner sighting through a bubble just behind the turret. It was found to be unsatisfactory and replaced with the familiar Sperry ball turret from the 113th B-17E onwards, this becoming a feature of subsequent combat versions. The ball turret was a claustrophobic's nightmare, being cramped and uncomfortable for the gunner housed within. It was only able to be entered (and exited) when the guns were pointed straight down and then only from the interior of the aircraft. The gunner would normally only get into the ball turret when necessary, but many were forced to make return trips crammed inside when the turret jammed after its hydraulic operating mechanism sustaining combat damage. Dangling from under the fuselage of the B-17 and unable to get out, a wheels up landing was not a pleasant occurrence for a ball turret gunner. Despite this, it was proven that statistically, this was one of the safest positions in the aircraft!

The additional defensive armament increased the B-17E's normal crew complement to 10 – two pilots, navigator, bombardier/nose gunner, ven-

Some nose detail of a B-17E. Visible are the single machine gun in the nose, the new dorsal turret and the windows in the cabin roof.

BOEING B-17 FLYING FORTRESS – PRODUCTION SUMMARY

Note: All B-17s up to and including the B-17E were built at Boeing's Seattle plant. Subsequent production was shared between Boeing (BO), Douglas at Long Beach (DL) and Vega at Burbank (VE). Production blocks numbered in increments of 5 except where noted.

Abbreviations: Qty – quantity; ff – first flight; del – delivery.

Model	Qty	USAAF Serials	Remarks
Model 299	1	–	ff 28/07/35, civil reg X-13372
Y1B-17	13	36-149/161	evaluation batch, del from 12/36
Y1B-17A	1	37-369	turbosupercharged engines, ff 29/04/38
B-17B	13	38-211/223	38-211 ff 27/06/39
B-17B	13	38-258/270	
B-17B	2	38-583/584	
B-17B	1	38-610	
B-17B	10	39-1/10	B-17B del completed 03/40
B-17C	38	40-2042/2079	ff 21/07/40, 20 to RAF 1941
B-17D	42	40-3059/3100	del 02-04/41
B-17E	277	41-2393/2669	ff 09/41
B-17E	235	41-9011/9245	
B-17F-BO	300	41-24340/24639	ff 05/42; Blocks 1-25, 27
B-17F-BO	435	42-5050/5484	Blocks 30-50
B-17F-BO	1565	42-29467/31031	Blocks 55-130
B-17F-DL	599	42-2964/3562	Blocks 1, 5-75
B-17F-DL	2	42-37714/37715	Block 80
B-17F-DL	4	42-37717/37720	Block 80
B-17F-VE	500	42-5705/6204	Blocks 1, 5-50
B-17G-BO	1085	42-31032/32116	ff 21/05/43; Blocks 1, 5-35
B-17G-BO	350	42-97058/97407	Blocks 40 and 45
B-17G-BO	600	42-102379/102978	Blocks 50-60
B-17G-BO	2000	43-37509/39508	Blocks 65-110
B-17G-DL	1	42-3563	Block 5
B-17G-DL	1	42-37716	Block 10
B-17G-DL	493	42-37721/38213	Blocks 10-30
B-17G-DL	250	42-106984/107233	Block 35
B-17G-DL	1000	44-6001/7000	Blocks 40-70
B-17G-DL	650	44-83236/83885	Blocks 75-95
B-17G-VE	300	42-39758/40057	Blocks 1, 5 and 10
B-17G-VE	600	42-97436/98035	Blocks 15-40
B-17G-VE	1000	44-8001/9000	Blocks 45-90
B-17G-VE	350	44-85492/85841	Blocks 95-110, last flown 29/07/45

PRODUCTION BY MANUFACTURER

	Boeing	Douglas	Vega	Totals
Model 299	1	–	–	1
Y1B-17	13	–	–	13
Y1B-17A	1	–	–	1
B-17B	39	–	–	39
B-17C	38	–	–	38
B-17D	42	–	–	42
B-17E	512	–	–	512
B-17F	2300	605	500	3405
B-17G	4035	2395	2250	8680
Totals	**6981**	**3000**	**2750**	**12731**

The B-17E production line in November 1941, two months after the first aircraft flew. The B-17E was the first Fortress variant to be built in quantity (512 produced) and the last solely manufactured by Boeing. (Boeing)

The two man cockpit of the B-17E. This is the first aircraft (41-2393) which recorded its maiden flight in September 1941. (Boeing)

tral turret gunner, flight engineer/dorsal turret gunner, tail gunner, two waist gunners and radio operator.

Other changes incorporated in the B-17E included the installation of windows in the cockpit roof, while additional manually operated defensive machine guns could be mounted in the nosecone and nose 'cheek' positions if necessary, plus in the original dorsal position immediately behind the bomb bay at the extreme aft end of the cockpit fairing in the radio operator's area.

The B-17E retained its predecessor's R-1820-65 Cyclone engines and despite being the heaviest of the series so far with a maximum weight of 53,000lb (24,040kg) it lost little speed compared with the B-17D, recording a maximum of 317mph (510km/h) at 25,000 feet.

Orders and Production

Orders for an initial quantity of 277 B-17Es were placed with Boeing in the second half of 1940, increasing to 812 shortly afterwards. Of these, 512 were completed as 'Es', the remainder becoming the first batch of B-17Fs. All of them built by Boeing at Seattle and the B-17E was the last Fortress variant to be produced solely by the parent company, Lockheed-Vega and Douglas making major contributions to the later marks' tallies. Plans for this expanded production base were put into place during 1940 at a time when the USA was still not at war and Pearl Harbour was more than a

year away. This foresight meant that when very large quantities of B-17s were needed the factories were there to provide them from 1942 onwards.

The first B-17E (41-2393) was flown on 5 September 1941, nearly five months behind schedule. The delay was caused by a shortage of materials resulting from the huge demands which were suddenly placed on US industry when massive orders

were taken for all types of military hardware as the need to supply Britain became more urgent and the US Government realised that it would also probably be drawn into the war.

The B-17E delay was quickly made up, the 512th and last aircraft being delivered ahead of schedule in May 1942. The average cost of a B-17E was just under $US300,000 and 45 were supplied to the RAF as the Fortress IIA under the terms of Lend-Lease.

The B-17E was the first truly combat worthy Fortress variant, entering service with the newly renamed US Army Air Force (USAAF) in the last couple of months of 1941 and seeing extensive service against Japan in the first 12 months of the Pacific war where it was used mainly as a tactical bomber with varying degrees of success. B-17Es were also sent to North Africa.

The Fortress's more conventional use as a high altitude strategic bomber would soon come in Europe, B-17Es of the 8th Air Force's 97th Bombardment Group in England recording the historic first use of American bombers against targets in Nazi occupied Europe on 17 August 1942. On that occasion, 12 97th BG B-17Es raided the marshalling yards at Rouen, the river port in north-western France. With this, the die was cast for the B-17's best known role.

The famous (or infamous!) Sperry ball turret, introduced to the Fortress with the 113th B-17E off the line in early 1942. The two 0.50in machine guns were each fed by 500 rounds of ammunition. (Boeing)

B-17E FORTRESS
Powerplants: Four Wright R-1830-65 Cyclone nine cylinder, single row, turbosupercharged radial engines rated at 1,200hp (895kW) for takeoff and 1,000hp (745kW) at 14,200ft; Hamilton Standard constant-speed and feathering three bladed propellers of 11ft 6in (3.50m) diameter; fuel capacity 1,730 US gal (6,549 l).
Dimensions: Wing span 103ft 9in (31.62m); length 73ft 10in (22.50m); height 19ft 2in (5.84m); wing area 1,420sq ft (131.9m²).
Weights: Empty 33,280lb (15,096kg); loaded 53,000lb (24,040kg).
Armament: Normal bomb load 4,000lb (1,814kg); eight 0.50in machine guns plus one 0.30in machine gun.
Performance: Max speed 275kt (510km/h) at 25,000ft; cruising speed 183kt (338km/h); time to 10,000ft 7.1min; service ceiling 36,600ft (11,155m); range (4,000lb bomb load) 1,740nm (3,218km); max range 2,780nm (5,150km).

B-17F FORTRESS

If the B-17E Fortress was the first variant to achieve quantity production, the B-17F was certainly the first to be subject to mass production with 3,405 examples built in a period of only 15 months during 1942 and 1943. The B-17F was manufactured at three different locations: Boeing Seattle (2,300 aircraft), Lockheed-Vega at Burbank California (500) and Douglas Long Beach California (605). Each cost an average $US358,000 and 19 were transferred to the RAF as the Fortress II.

The build up of B-17F production from Vega and Douglas was at first slow with only 68 and 85 aircraft, respectively, having been turned out by the end of 1942, compared with Boeing's 850. The rate quickly built up during 1943 and by the time the B-17F was giving way to the 'G' on the pro-

duction lines in mid 1943, the three factories were turning out 400 Fortresses each month.

The first B-17F (41-24340) was flown in May 1942 and externally was very similar to the B-17E, the main difference being the addition of a slightly longer, unframed plexiglass nose cone. Under the skin there were more than 400 changes, however, including the installation of R-1820-97 engines which produced the same basic 1,200hp (895kW) but were capable of a 1,380hp (1,030kW) 'war emergency' rating for a limited period.

Paddle bladed propellers of slightly greater diameter were fitted; the engine cowlings with reshaped more rounded leading edges (to give the new propeller blades more space when feathered); self sealing oil tanks were installed along with an improved oxygen system; the control layout was improved; the bomb racks changed; wings and undercarriage strengthened to cope with ever increasing weight; a link between the automatic pilot and the Norden bombsight was added and so on.

Running changes introduced to the B-17F production line included extra fuel cells in the outer wings (the so-called 'Tokyo' tanks) bringing capacity to 2,810 USgal (10,637 l), and the ability to carry four additional 1,000lb (454kg) bombs externally on racks under the wings.

Ever increasing weight is a fact of life for most aircraft, but in the case of the difference between the B-17F and its predecessor the gap was substantial, maximum weight increasing from 53,000lb (24,040kg) by no less than 23 per cent to 65,500lb (29,711kg).

The basic defensive armament fit remained the same as the B-17E with various options available for the installation of socket mounted machine guns in the nose and 'cheek' areas

plus the dorsal gun which could be manned from the radio operator's compartment. Thus it was possible for a B-17F to have up to 13 machine guns installed, all of them now of the 50-calibre variety.

A substantial modification was made to late production Douglas built B-17Fs, the last 86 of which incorporated the Bendix chin turret more normally associated with the B-17G. Housing twin 'fifties', this powered turret substantially increased the Fortress's protection from a frontal attack, an area in which it had previously been vulnerable. The turret was subsequently fitted to a number of earlier B-17Fs in the field.

B-17F FORTRESS
Powerplants: Four Wright R-1820-97 Cyclone nine cylinder, single row, turbosupercharged radial engines rated at 1,200hp (895kW) for takeoff and 1,380hp (1,030kW) war emergency power at 25,000ft; Hamilton Standard constant-speed and feathering three bladed propellers of 11ft 7in (3.53m) diameter; fuel capacity 1,730 US gal (6,549 l) or 2,810 USgal (10,637 l).
Dimensions: Wing span 103ft 9in (31.62m); length 74ft 9in (22.78m); height 19ft 2in (5.84m); wing area 1,420sq ft (131.9m²).
Weights: Empty 34,000lb (15,422kg); max loaded 65,500lb (29,711kg).
Armament: Normal bomb load 4,000lb (1,814kg), normal maximum 8,000lb (3,629kg); 10-13 0.50in machine guns.
Performance: Max speed (war emergency power) 273kt (505km/h) at 25,000ft; max speed (normal power) 260kt (482km/h); cruising speed 174kt (322km/h) at 10,000ft; time to 20,000ft 25.7min; service ceiling 37,500ft (11.430m); range with 4,000lb bombs 1,914nm (3,540km); range with 6,000lb bombs 1,130nm (2,092km); max fuel range (1,730 USgal) 2,330nm (4,313km); max fuel range (2,810 USgal) 3,300nm (6,115km).

A Boeing built B-17F-95-BO. The F's main external change over its predecessor was the fitting of a lengthened, frameless plexiglass nosecone. Inside, there were more than 400 modifications including the operationally important one of linking the aircraft's Norden bombsight and autopilot together. (Boeing)

The B-17F's new paddle bladed propellers of slightly greater diameter are well illustrated in this ground shot of an aircraft undergoing maintenance. The F was the first mass produced Fortress, 3,405 coming from Boeing, Douglas and Lockheed-Vega production lines.

B-17G FORTRESS

By far the most numerous of all the Fortress variants, the definitive B-17G accounted for 68 per cent of production with no fewer than 8,680 coming off the Boeing (4,035), Vega (2,250) and Douglas (2,395) production lines between 1943 and 1945 at an average rate of more than 330 per month. The first B-17G (Boeing built 42-31032) was flown on 21 May 1943 and the last (Vega built 44-85841) on 29 July 1945. Production at Boeing and Douglas had phased out a couple of months earlier. The RAF received 85 as the Fortress III.

With mass production came greater efficiencies. Where it took 54,800 man-hours to assemble a B-17E in 1942 it took just 18,600 man-hours to build the more complex B-17G in 1944. This was reflected in the price of the aircraft. Where as the average B-17E cost $US298,000, a B-17G built in 1944 cost $US204,370 and in 1945 the figure was further reduced to $US187,742. The peak production rate achieved by all three factories combined was 130 per week.

The B-17G was fundamentally similar to later production B-17Fs but with the addition of the Bendix chin turret which had appeared on a few Douglas built examples of the earlier model. Some very early B-17Gs lacked this feature due to the change-over of production but all had the F's extra fuel tanks, Sperry ball turret and in all but a very few examples, R-1820-97 engines with their 1,380hp (1,030kW) 'war emergency' rating.

The powerplants received some upgrading early in the production run with the fitting of electrically operated turbosupercharger regulators instead of the original mechanical units. These were much more reliable than the originals and also more precise in

The B-17G accounted for more than two-thirds of all Fortress production with 8,680 emerging from three factories. Illustrated is a Boeing built B-17G-45-BO with the new Bendix chin turret fitted to this model displayed for the camera. (Boeing)

B-17Gs in production. At its peak, combined Boeing, Douglas and Lockheed-Vega B-17G production reached 130 per week. It took only 18,600 man-hours to build a B-17G in 1944 compared with nearly three times that for a B-17E two years earlier. These Boeing production line shots were taken in December 1943. (Boeing)

Late production B-17Gs from Douglas and Vega with some variations in defensive armament evident. The lower and centre aircraft are lacking tail and dorsal guns while the centre example is also missing its chin turret, as is the top aircraft. All three have (or did have) the Cheyenne rear turret installed, identifiable by the gunner's greater glass area and shorter gun mounting section. The aircraft are (from nearest the camera): B-17G-95-DL 44-83514; B-17G-70-VE 44-8543 and B-17G-95-DL 44-83872, the 15th last Fortress built by Douglas out of 3,000.

operation, the mechanical regulators sometimes making formation flying difficult. The B-17G also received improved General Electric turbosuperchargers in 1944, giving better high altitude performance and an improvement in the aircraft's critical altitude.

Modifications

Other changes were made during the course of production, most of them associated with defensive armament and operational equipment. The first occurred in late 1943 when the Sperry dorsal turret was replaced with a more bulbous Bendix unit which provided better visibility and control; the waist gun positions were staggered in order to give the gunners more room to move (the starboard position was moved forward) and various covers were introduced to help protect the gunners from the slipstream; and the radio room manually operated gun in the top of the fuselage received a revised installation which enabled the area to remain screened when the gun was fitted.

Combat experience over Europe eventually saw the removal of this gun from most aircraft as its effectiveness was questionable and with the ever increasing weight of the aircraft, cutting down on the amount of ammunition and other equipment carried became important. Reducing fighter opposition in the final stages of the war also made this gun less than necessary. The cheek guns' in-

stallation varied slightly as it had with the B-17F, some guns being mounted in a very forward position just behind the nose cone and others a short distance further aft. Small shields to offer protection from the slipstream were also installed.

An important modification was made to the tail gun with the installation of a new turret design. Called the Cheyenne after the United Air Lines modification centre in Wyoming where it was developed, the turret offered greater gun elevation and a reflector gunsight instead of the previous ring and bead system. The tail gunner sat in an enclosure with greater glass area and therefore visibility and the guns' operating mech-

anism was improved. The guns themselves were housed in a pivoting cupola which was closer to the gunner, reducing the overall length of the aircraft by five inches (13cm).

The increasing weight of the B-17G caused handling problems and increased the strain on the Wright Cyclone engines. In reality, the B-17G could have done with more power at a time when normal maximum weight was 65,500lb (29,711kg) and a maximum overload of 72,000lb (32,659kg) was available. This represented a 45 per cent increase over the B-17C and 35 per cent over the B-17E, both of which had the same 1,200hp (895kW) available for takeoff from each of the engines. By comparison, the Lancaster had a similar maximum weight but its Merlin 24 engines each produced 1,620hp for takeoff.

Weight saving therefore became a priority and as German fighter resistance diminished, many B-17Gs flew on missions without waist guns and in some cases even without chin and ball turrets. Armour protection for the crew was also often removed but so called 'flak curtains' (laminated plates in canvas to protect against shell splinters) had to be fitted as anti aircraft gun activity increased. These modifications reduced weight substantially and increased maximum speed by up to 20mph (32km/h) with commensurate improvements in payload/range, climb and ceiling performance.

There were also some internal modifications to the B-17G in an attempt to ease the cramped conditions under which the crew operated. The navigator received a larger table with a map shelf above it and a swivelling chair while various items of equipment were rearranged.

A welcome addition was the so called 'formation stick' to help fine control of the heavily laden aircraft at high altitude when flying in tight for-

Vega built B-17G-110-VE 44-85818 from the final production batch, the last Fortress of which was flown on 29 July 1945. This aircraft is shown with its H2X 'Mickey' radar radome extended. This was installed in place of the usual ball turret.

Completed B-17Gs everywhere at Boeing's Seattle plant in 1944. These aircraft are from the Block 65 production batch which covered the serial numbers 43-37509 to 43-37673. (Boeing)

mation. This was basically as electrically operated power boost for the control column, operated by a pistol grip device attached to the column. It appeared on production aircraft in early 1945 and considerably eased the pilot's workload once a formation was established. The improved turbosupercharger regulator mentioned earlier also helped in this area.

Radar Bombing

Although the B-17 was used almost entirely for day bombing missions, its effectiveness was often compromised when the target was covered by cloud with the result that many missions had to be aborted or secondary targets attacked instead. Radar was the obvious answer and the first tests were conducted by the 8th Air Force in the second half of 1943 using B-17Fs with British ground mapping H2S radar (as was fitted to RAF Lancasters) installed in a blister radome under the nose. These Fortresses were used as pathfinders for the main formations, signalling when bombs should be dropped.

An American development of H2S – called H2X – entered service in November 1943, initially fitted to 11 Douglas built B-17Gs and installed in a partially retractable plastic dome immediately behind the chin turret. This caused the forward compartment of the Fortress to become very

cramped, resulting in the equipment being moved aft into a retractable radome in the place once occupied by the ball turret using the same mechanism to raise and lower it. The radar operator was housed in the radio compartment behind the bomb bay.

Officially called Bombing Through Overcast (BTO) radar, the equipment was quickly nicknamed 'Mickey' radar and B-17s so equipped were assigned initially to specialist pathfinder units and then to normal bomber

groups, most of which had a few 'Mickeys' on strength for leading missions.

The USAAF also dabbled with some of the navigation and bombing systems used by the RAF including Oboe and Gee-H (refer Lancaster section), the latter extensively used in the last few months of the European War, while some B-17Gs were equipped with AN/APQ-7 'Eagle' ground mapping radar which gave a much clearer display than the older H2S/H2X.

B-17G FORTRESS

Powerplants: *Four Wright R-1820-97 nine cylinder, single row turbosupercharged radial engines rated at 1,200hp (895kW) for takeoff and 1,380hp (1,030kW) war emergency power at 25,000ft; Hamilton Standard constant-speed and feathering three bladed propellers of 11ft 7in (3.53m) diameter; fuel capacity 2,810 USgal (10,637 l), provision for 820 USgal (3,104 l) overload tanks in bomb bay.*

Dimensions: *Wing span 103ft 9in (31.62m); length (standard rear turret) 74ft 9in (22.78m); length (Cheyenne rear turret) 74ft 4in (22.66m); height 19ft 1in (5.82m); wing area 1,420sq ft (131.9m²); tailplane span 43ft 0in (13.10m).*

Weights: *Basic empty 36,135lb (16,390kg); empty equipped 38,000lb (17,237kg); normal max takeoff 65,500lb (29,711kg); max overload 72,000lb (32,659kg).*

Armament: *Normal bomb load (long range mission) 4,000lb (1,814kg); normal max 9,600lb (4,354kg); 10 to 13 0.50in machine guns in chin (365 rpg), dorsal (375 rpg), ventral (500 rpg), tail (500 rpg), waist (600 rpg), radio compartment and cheek (310 rpg) positions*

Performance: *Max speed (war emergency power) 263kt (486km/h) at 25,000ft; normal max speed 250kt (462km/h) at 25,000ft; normal cruise 158kt (293km/h) at 10,000ft; economical cruise 139kt (257km/h); time to 20,000ft 37.0min; service ceiling 35,600ft (10,850m); takeoff distance 3,400ft (1,036m); range with 4,000lb bombs 1,565nm (2,900km) at 25,000ft; max range (no load, standard fuel) 2,958nm (5,471km); max range (no load, auxiliary fuel) 3,820nm (7,080km).*

The first of the armed bomber escort conversions, the XB-40, converted from an early B-17F. 'Production' YB-40 conversions had up to 16 0.50in machine guns.

(above) Fuselage detail of a YB-40 showing the twin gun waist position installation and the second, mid fuselage, dorsal turret.

FORTRESS CONVERSIONS

Many new Fortress variants were created by conversion to trainers, transports, drones, bomber escorts, search and rescue, reconnaissance aircraft and others both during World War II and afterwards. The following lists most of them:

XB/YB-40: Developed as a heavily armed bomber escort intended to fly with the normal bomber formations, the YB-40 was equipped with extra guns and a vast supply of ammunition. All except the first prototype were converted from Vega built B-17Fs.

The first aircraft (41-424341, the second Boeing built B-17F) was flown in early 1943 as the XB-40. Twenty YB-40 and four TB-40 (trainer)

(below) One of about 50 B-17H 'Dumbo' air-sea rescue Fortresses converted from B-17Gs. The ground crew is photographed practising installing the aircraft's droppable lifeboat. (Boeing)

The sole XB-38 conversion with Allison V-1710-89 engines survived only one month before suffering an engine fire. (Boeing)

conversions followed in the same year, the aircraft's impressive machine gun armament sitting in the usual cheek, tail, dorsal and ball turret installations plus the newly developed chin turret, two more in a second dorsal turret at the radio operator's station behind the bomb bay and twin installations in the waist positions instead of the usual single guns. The result was a gun armament of up to 16 0.50in Brownings fed by over 11,000 rounds of ammunition, much of which was stored in the bomb bay.

The YB-40s were deployed to the 327th Bombardment Squadron at Alconbury, England, in May 1943 and flown on just nine missions between then and the end of July when the idea was abandoned. They were not completely successful due to poor flying characteristics resulting from an awkward rearwards biased centre of gravity which made formation flying difficult. Moreover, once the normal bomber B-17s had dropped their payload they were considerably lighter than the YB-40 and therefore faster. The result was that the bombers had to reduce power in order to let the YB-40s stay with them, a less than desirable situation over enemy territory.

Serial numbers were: XB-40 – 41-24341; YB-40 – 42-5732/5744, 42-5871, 42-5920/21, 42-5923/25, 42-5927; TB-40 – 42-5833/34, 42-5872, 42-5926.

XB-38: B-17E 41-2401 re-engined with four 1,425hp (1,065kW) Allison V-1710-89 liquid cooled inline engines. First flown in its new guise in May 1943, the aircraft was lost just one month later after an engine fire before comparative trails with a standard B-17 could be completed.

B-17H/SB-17: About 50 B-17Gs converted for search and rescue duties in 1945 were designated B-17Hs and nicknamed 'Dumbo'. Carrying an underslung droppable lifeboat, the aircraft were used in the North Sea area and from Iwo Jima in the Pacific covering the B-29 Superfortress operations. The B-17Hs were redesignated as SB-17Gs in 1948 while aircraft operated by the US Coast Guard were given the designation PB-1G. The final regular US Air Force unit to operate the Fortress was the 57th Air Rescue Squadron based in the Azores which retired its last SB-17 in 1956.

A US Coast Guard PB-1G Fortress, 17 of which were operated for air-sea rescue and iceberg reconnaissance duties. (Boeing)

(below) A QB-17 drone converted from a B-17G. This aircraft completed a 2,600 miles (4,200km) crewless non stop flight from Hawaii to the mainland USA in August 1946.

PB-1W/PB-1G: The US Navy acquired 48 B-17Gs in 1945 of which 31 were equipped with APS-20 search radar mounted in an underfuselage radome for anti submarine and early warning patrol duties and designated PB-1Ws. The remaining 17 aircraft were transferred to the US Coast Guard for air-sea-rescue (with underslung lifeboat) and iceberg reconnaissance duties under the designation PB-1G. The last Coast Guard PB-1G (77254, ex USAAF 44-85828) was the final manned B-17 variant to serve with any of the US forces, flying its ultimate mission in October 1959.

QB-17M/N: Designation applied to radio controlled Fortress conversions, controlled by operators either on the ground or in DB-17 director aircraft (see below). QB-17s are perhaps best known for their use in the Bikini Atoll atomic tests in 1946-47 to test the effects of radiation and blast turbulence, and as targets for various air-to-air and air-to-ground missiles.

DB-17P: Drone director aircraft converted from B-17Gs to control QB-17s (above).

RB-17: A dual designation, RB-17 was originally applied to B-17B/C/D aircraft which had been withdrawn from regular service, the 'R' standing for 'restricted'.

The subsequent RB-17 designation applied to reconnaissance versions of the Fortress, formerly dubbed F-9 or FB-17 (see below). An RB-17G flew the first mission of the Korean War in June 1950, a photo mapping sortie over North Korea.

F-9: The original designation applied to Fortresses converted for photographic reconnaissance work. The F-9, F-9A and F-9B were conversions of B-17Fs (41 converted) fitted with six cameras while the F-9C (10 conversions) was based on the B-17G.

TB-17: A loose designation applied to any B-17 version used for training.

VB-17G/CB-17: The VB-17 was a postwar unarmed VIP transport with all military equipment removed and a plush interior installed including office and sleeping accommodation. Less sumptuously appointed (but still very comfortable) staff transport conversions were dubbed CB-17s. Several served until well into the 1950s.

MB-17: Designation applied to a small number of aircraft converted for use as launch platforms for the testing of various guided missiles.

JB-17G: (Boeing Model 299Z) Two postwar conversions as engine testbeds. The two aircraft (from B-17Gs 44-85813 and 44-85734) were originally called EB-17Gs, the designation later changing to JB-17G. Both were originally intended to test massive new turboprop engines with these powerplants installed in the nose. The original Wright Cyclone piston engines were retained.

The first JB-17G went to Curtiss-Wright for installation of the 5,500ehp (4,105kW) XT-35 Typhoon turboprop while the second was operated by Pratt & Whitney to test its similarly rated XT-34 engine. Both were subsequently fitted with other engines including the J65 turbojet (licence built Armstrong Siddeley Sapphire) mounted under the nose of the Wright aircraft.

Other Fortresses were also used as engine testbeds including by Allison which used one for its T56 turboprop which subsequently powered the Lockheed C-130 Hercules, P-3 Orion and Electra airliner.

One of two B-17Gs converted to engine testbeds under the designation JB-17G. Both were modified to carry the massive Curtiss-Wright XT-35 or Pratt & Whitney XT-34 turboprops in the nose. This is the XT-34 version with the civil registration N5111N. (via Philip J Birtles)

XC-108: A B-17E (41-2593) converted to a personal transport for General Douglas MacArthur in 1943 with armament and armour deleted, the bomb bay sealed and used as a storage compartment, windows added, a five seat passenger compartment installed in the forward waist area, galley and refrigeration facilities in the radio room area (radio equipment was moved to the rear of the cockpit) and an air-stair door. A similar B-17F (42-6036) conversion was performed for use in Europe and the Middle East, this aircraft being dubbed the YC-108.

XC-108A: A one-off conversion of a B-17E (41-2595) for cargo work with a stripped interior and upward opening freight door installed in the port fuselage just behind the wing. Numerous other B-17s were used for transport work, notably stripped down B-17Es used to carry supplies between Australia and New Guinea. These and other similarly converted B-17s used in Europe retained their original designations.

XC-108B: Another one-off conversion of a B-17F, this time as an experimental flying fuel carrier for operations over 'The Hump' route between Burma and China. Fuel cells installed in the bomb bay allowed 1,000 USgals (3,785 l) to be carried, but the Fortress's restricted fuselage space was not entirely suited for either this or the straight cargo carrying role.

BQ-7: A retrospective designation applied to war weary B-17Fs and Gs in a highly secret wartime project using radio controlled Fortresses as unmanned flying bombs. Codenamed *Aphrodite*, the project involved stripping the aircraft of all guns, armour and bombing equipment and then packing the it with 20,000lb (9,072kg) of high explosive. About 25 aircraft were converted.

A two man crew flew the aircraft to the appropriate altitude and course whereupon they bailed out and control was taken over by an accompanying radio 'mother ship' which guided the aircraft to its target. Trials and operations took place in the second half of 1944 and were terminated in early 1945 after it was realised that the chances of hitting the selected target were slim due to the limitations and unreliability of the early radio control equipment. One BQ-7 crash landed in Germany without exploding, revealing its secrets to the enemy while others came down out of control in or near England with spectacular if unfortunate results.

Of interest is the fact that Joseph P Kennedy Jr (brother of future US President John Kennedy) was killed in a BQ-7 when it exploded before he and his fellow crew member could bail out. The fact that one of the famous Kennedy family was involved generated considerable interest in US government circles and undoubtedly hastened the end of the programme.

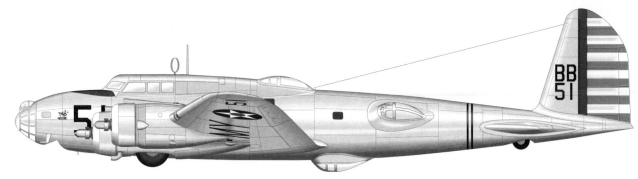

Y1B-17 Fortress of 2nd BG/20th BS USAAC.

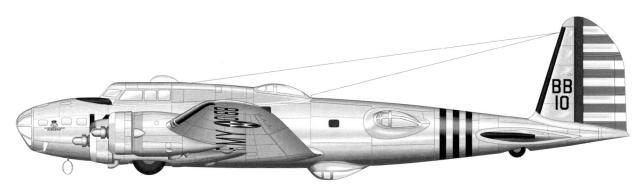

B-17B Fortress of 2nd BG USAAC.

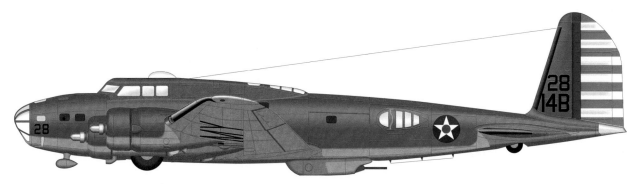

B-17C Fortress of 19th BG/14th BS USAAC, Philippines 1941.

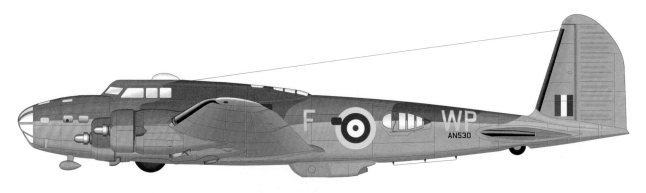

B-17C Fortress I AN530/WP-F (formerly USAAC 40-266) of 90 Squadron RAF late 1941.

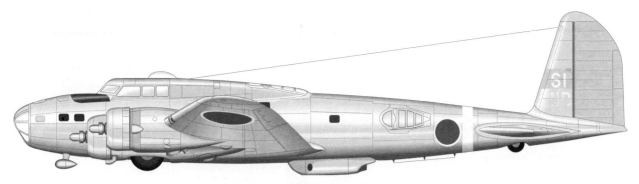

B-17D Fortress captured by the Japanese at Clark Field, the Philippines. Aircraft built up from several 19th BG aircraft damaged by a Japanese raid on the airfield on 8 December 1941 and tested in Japan by the Japanese Army.

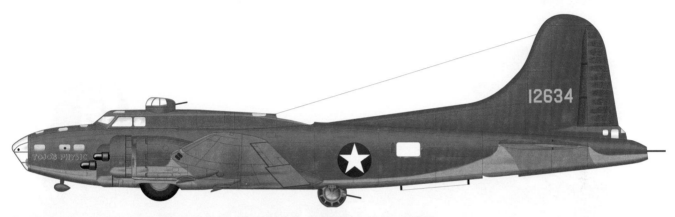

B-17E Fortress 41-2634 'Tojo's Physic' of 19th BG USAAF, South West Pacific Area 1942.

B-17F-BO Fortress 41-24485 'Memphis Belle' of the 91st BG/324th BS USAAF, England 1943.

B-17F-BO Fortress 41-24554 'The Mustang' of the 19th BG USAAF, Pacific 1943. Mission and kill markings: 109 missions, 2 submarines, 43 small cargo ships, 3 enemy vessels and 17 aircraft.

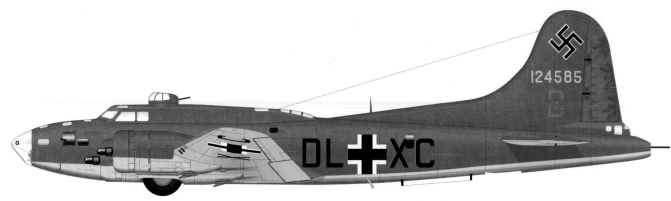

B-17F-BO Fortress 41-24585 originally of 303rd BG USAAF. Forced down during mission to Rouen marshalling yards, France, on 12 December 1942. Made wheels down landing in field, captured by Germans and operated by Kampfgeschwader 200 (KG200).

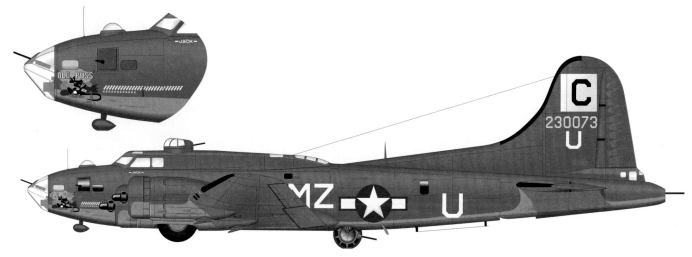

B-17F-BO Fortress 42-30073 'Ole Puss' of 96th BG/413th BS, England 1943.

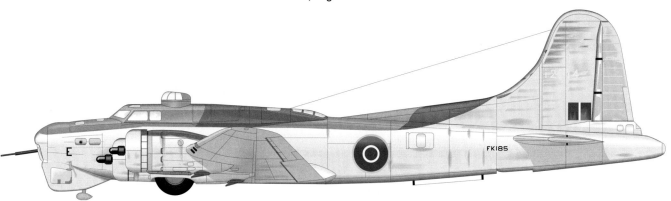

B-17E Fortress IIA FK185 (ex USAAF 41-2514) of 220 Squadron RAF Coastal Command 1942. Fitted with remotely sighted 40mm Vickers 'S' cannon in nose for anti shipping and anti submarine use.

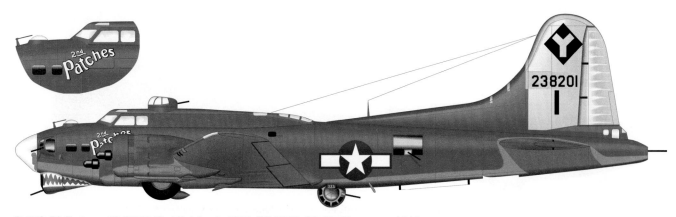

B-17G-DL Fortress 42-38201 '2nd Patches' of 99th BG/346th BS, Mediterranean 1944.

B-17G-BO Fortress 42-39163 of 486th BG/835th BS, England late 1944. Lost during a raid on 7 April 1945. Note Cheyenne rear turret.

B-17G-DL Fortress 44-6884 'Kwiturbitchin' of 97th BG/414th BS, Mediterranean 1945. Striped rudder denotes squadron lead aircraft.

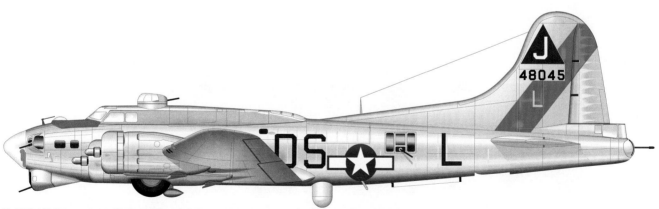

B-17G-VE Fortress 44-8045 of 351st BG/511th BS, England early 1945, with H2X radar radome extended.

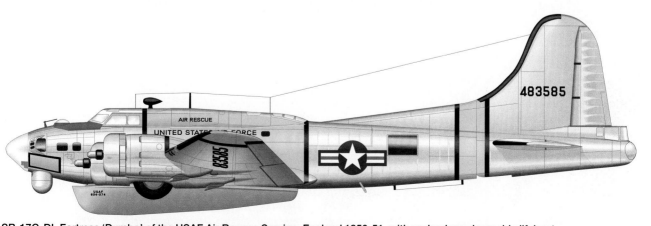

SB-17G-DL Fortress 'Dumbo' of the USAF Air Rescue Service, England 1950-51, with underslung droppable lifeboat.

B-17 FLYING FORTRESS AT WAR

RAF FORTRESSES

The first Fortresses to see combat where the 20 B-17Cs delivered to Britain's Royal Air Force in 1941 as the Fortress I. Serialled AN518 to AN537, these aircraft were operated by No 90 Squadron operating from bases in Norfolk and were flown on combat missions for a brief period before being assigned to other roles.

The Fortresses were diverted to Britain following representations to the US Government by the British Purchasing Commission in the second half of 1940. Approval was given late in the year after considerable discussion as the Americans were wary of their intended use, suspecting the RAF wanted to use them on raids to Berlin. This, of course, was perfectly true! It had initially been agreed that the first six aircraft would go to Coastal Command but Britain's intended use of the other 14 remained

vague. In the end, all 20 went to Bomber Command.

Considering that 20 B-17Cs represented a large proportion of total Fortress production by late 1940 (only about 90 of all models including the prototype and evaluation batch had been built by then), the USA's decision to release them to the RAF could be considered generous, although there was a certain amount of benefit to the Americans as well because invaluable information gathered from combat experience would be passed on to them.

Combat quickly highlighted the B-17's weaknesses – inadequate defensive armament, lack of proper protection for the fuel tanks, lack of armour protection for the crew, the installation of a Sperry rather than the still secret Norden bombsight and poor bomb load by British standards. Despite these concerns – which were

appreciated in advance anyway – there were some pluses for the British in that the B-17 would provide the RAF with a genuinely high altitude bomber and one which could easily reach Berlin.

The latter was an important political point championed by Prime Minister Winston Churchill, who saw the B-17s as surrogate US bombers for use in attacks on Germany. This was all part of Churchill's desire to get America involved in the war in whatever way he could, officially or otherwise.

The altitude performance of the Fortress was far and away better than any other RAF bomber and required special training for the crews who would fly it. They had to undergo a four hour test at a simulated 35,000 feet (10,668m) in a decompression chamber breathing oxygen and emerging without ill-effect in order to be passed as fit to fly in the B-17.

High altitude, daylight precision bombing as practised by the B-17 Fortresses of the 8th and 15th Air Forces in Europe. (Boeing)

44-85784 'Sally B', a Fortress which remains airworthy today. She was photographed here in 1977. (Philip J Birtles)

Deliveries and Action

Delivery of the 20 Fortress Is was carried out in April and May 1941, the first aircraft (AN521) arriving in Britain after a record 8hr 26min flight across the Atlantic. The Fortresses all had incorrect serial numbers applied in the USA, with 'AM' instead of the correct 'AN' prefixes painted on.

The need to incorporate modifications and to perform evaluation and training sorties caused some delays in 90 Squadron working up to operational status, this not being achieved until early July. The Fortress was something of an orphan in RAF service and required special provisioning. Apart from the obvious problems with spare parts, it could also only carry American bombs. Its performance characteristics also dictated that 90 Squadron's operations would be carried out in isolation rather than as part of combined operations with other squadrons and aircraft types.

Two Fortresses were lost before operations began, one due to encountering heavy icing at 35,000 feet and then extreme turbulence, the combination resulting in a terminal velocity dive and the breakup of the airframe; and the other when an engine caught fire during ground running, the uncontrolled blaze destroying the aircraft.

The very first B-17 combat mission took place on 8 July 1941 when three of 90 Squadron's aircraft were sent to attack the submarine building docks at Wilhelmshaven. Each Fortress was armed with four 1,100lb (500kg) demolition bombs and the attack was to take place from a height of 27,000 feet (8,230m). Two aircraft managed to put their bombs on the target while the third was forced to abort due to oil coming out of the breathers of all four engines and then freezing on the tailplane, causing severe vibrations.

The other two Fortresses climbed to 34,000 feet (10,360m) after attacking the target, resulting in more oil throwing problems and considerable discomfort for the crews. An attempted interception by two Messerschmitt Bf 109Es caused no serious problems but the Fortresses were unable to return fire as the navigator/fire controller's windows were frosted over, the guns were frozen and the intercom system was inadequate to the task of proper communication.

The difficulties associated with high altitude flying had been rammed home, but to operate the aircraft at lower altitudes in daylight would have been even more dangerous as the Fortress's poor defensive armament would have offered scant protection against German fighters.

90 Squadron's planned second mission with the B-17s was against Berlin on 23 July. This time three aircraft were sent out but two had to quickly abort because the contrails they were leaving behind advertised their whereabouts to the *Luftwaffe* all too boldly, while the third also had to abort because it couldn't maintain height.

The Fortresses attacked various other targets between then and late September 1941 when operations were abandoned. By that stage only 11 of the original 20 aircraft remained due to combat losses and to other crashes. A raid on Oslo to attack the pocket battleship *Admiral Scheer* on 8 September sealed the B-17's fate as an operational bomber in RAF service. Four aircraft set out to attack the ship of which only two returned. One disappeared without trace and the other shot down by two Bf 109Fs.

Neither of the other two Fortresses managed to hit the target, one abandoning the attack because of cloud over the target area and the other jettisoning its bombs in order to escape the attentions of enemy fighters. Although badly damaged, this Fortress managed to get back to Kinloss in Scotland where it crash landed and was written off.

Four more individual aircraft sorties were flown between then and 25

RAF B-17C Fortress Is lined up at the Boeing factory prior to their trans-Atlantic delivery flights in April and May 1941. Note the incorrect 'AM' serial prefixes. (Boeing)

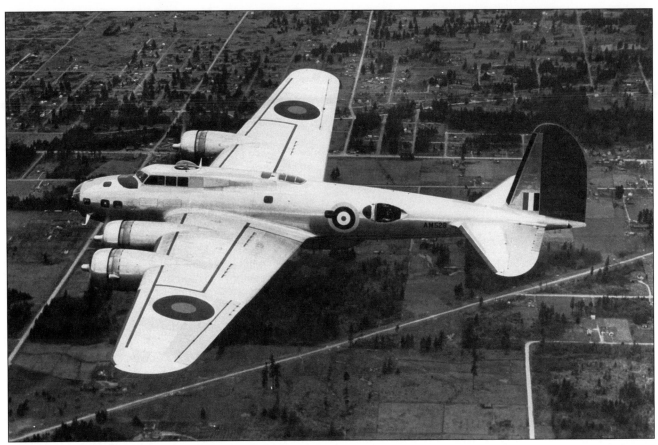

RAF Fortress I 'AM528' (actually AN528) before delivery. This aircraft was destroyed by a ground fire in July 1941, just a few weeks after delivery and before it had a chance to see combat. (Boeing)

September, after which the Fortresses were withdrawn from Bomber Command's operational fleet in Europe. Four of the survivors were sent to the Middle East for operational trials and in late January 1942 all the aircraft were transferred to Coastal Command serving with No 220 Squadron in Northern Ireland pending the arrival of B-17E Fortress IIs (the early aircraft continued in service as operational trainers after that) and No 206 Squadron.

The RAF's experience with the Fortress had been far from satisfactory and the service's damning reports on the aircraft caused some consternation on the other side of the Atlantic. Despite this, most of the criticisms were quickly acted on by both the US military and Boeing. At this point of its development, the B-17 was not a very good bomber, although the way in which the RAF operated it in the second half of 1941 was not overly conducive to success.

The RAF also operated later model Fortresses, receiving 200 B-17Es, Fs and Gs (as the Fortress IIA, II and III, respectively) mainly for service with Coastal Command on anti submarine patrols. Some B-17Gs were also used for high altitude weather reconnaissance and radio countermeasures duties, the latter in support of RAF bomber operations. These Fortresses carried jamming devices to saturate German radar screens during night bombing missions, while some of the anti submarine aircraft were fitted with a single 40mm Vickers 'S' cannon in the nose for attacking U-boats and other vessels on the surface.

THE PACIFIC CAMPAIGN

Although the Pacific War with Japan is best known as being the do-

An RAF B-17E Fortress IIA (foreground) before delivery carrying a combination of US and British markings. (Boeing)

main of the Consolidated B-24 Liberator and Boeing B-29 Superfortress as far as USAAF heavy bombers are concerned, the B-17 Fortress was heavily involved in the early part of this campaign, including its very first minutes when Japan launched its attack on the US naval base at Pearl Harbour, Hawaii on the morning of 7 December 1941.

Some say the B-17 did not come to the war in the Pacific, rather that the war came to it. As part of its gradual buildup during 1941, the USAAF had a very small force of just 35 B-17Cs and Ds in the area at the time, 12 aircraft of the 5th Bombardment Group at Hickham Field, Hawaii and 23 of the 7th and 19th BGs based at Clark Field in the Philippines. Twelve unarmed 7th BG Fortresses arrived at Hickham on 7 December, while the Japanese raid was in progress!

As a result of all this, the B-17C and D models were the first Fortresses to see action in USAAF service. The Japanese raids on Pearl Harbour and on the Philippines just nine hours later took a heavy toll on the B-17s. Five of the 12 aircraft on the ground at Hickham were destroyed, as was one of the arriving aircraft, while the Philippines attacks cost another 14 B-17s.

The survivors of the USAAF's B-17 force in the Pacific rallied to perform the aircraft's first offensive combat mission in US colours on 10 December when five 19th BG aircraft attacked a Japanese invasion convoy off the coast of Luzon. These were the first bombs dropped by an aircraft carrying US markings in World War II and although several hits on the ships were reported, none was

After deciding the Fortress had severe limitations as a bomber, the RAF put its aircraft to work with Coastal Command, mainly on anti submarine patrols. This B-17E Fortress IIA (FL459) flew with No 206 Squadron in that role, equipped with ASV radar. (via Neil Mackenzie)

sunk. Some more attacks on shipping followed over the next few days.

Withdrawal and Reorganisation

After this unsuccessful series of raids, the remaining 10 combat worthy B-17s were withdrawn to Australia from where they were quickly deployed to Java in an attempt to stop the Japanese overrunning the Netherlands East Indies. In the meantime, reinforcements in the shape of B-17Es were sent to the area, some 80 Fortresses of all versions being involved in the battle before the NEI campaign was lost in March 1942.

Of these, no fewer than 58 were lost to enemy action and accidents, including 19 on the ground in Java. The B-17s flew 350 missions in just over two months during the NEI campaign but the return on the investment in men and aircraft was poor with just two Japanese ships sunk. One of the problems facing the B-17 units was the fact that the aircraft was designed for strategic offensive strikes but circumstances dictated it be used for defensive tactical missions.

The setbacks resulted in the

USAAF B-17 force once again retreating to Australia, reorganising, establishing new units in the area and receiving new aircraft. A small pause also gave the air and ground crews a little time to organise themselves as they were all inexperienced and facing a lack of proper maintenance and organisational support.

The reorganisation of the units in the area saw the 7th BG redeployed to India to help stop the Japanese advance in that part of the world while the 4th, 11th, 19th and 43rd BGs (each with a strength of four 17 aircraft squadrons) remained in the Pacific. The B-17E became the standard equipment of these units and it was this model which bore the brunt of the USAAF's bombing responsibilities in the Pacific until the B-24 Liberator began to take over.

By the time the B-17 forces were better organised and equipped the Japanese advances had been all but halted, the decisive battles of the Coral Sea (May 1942) and Midway (June 1942) being fought mainly by US Navy aircraft. The B-17 did have a small involvement at Midway but

Disaster at Pearl Harbour on 7 December 1941. Apart from the losses sustained by the US Navy, the USAAF lost 152 out of 231 aircraft including 12 B-17s. This shot shows some B-17s burning at Hickham Field after the attack.

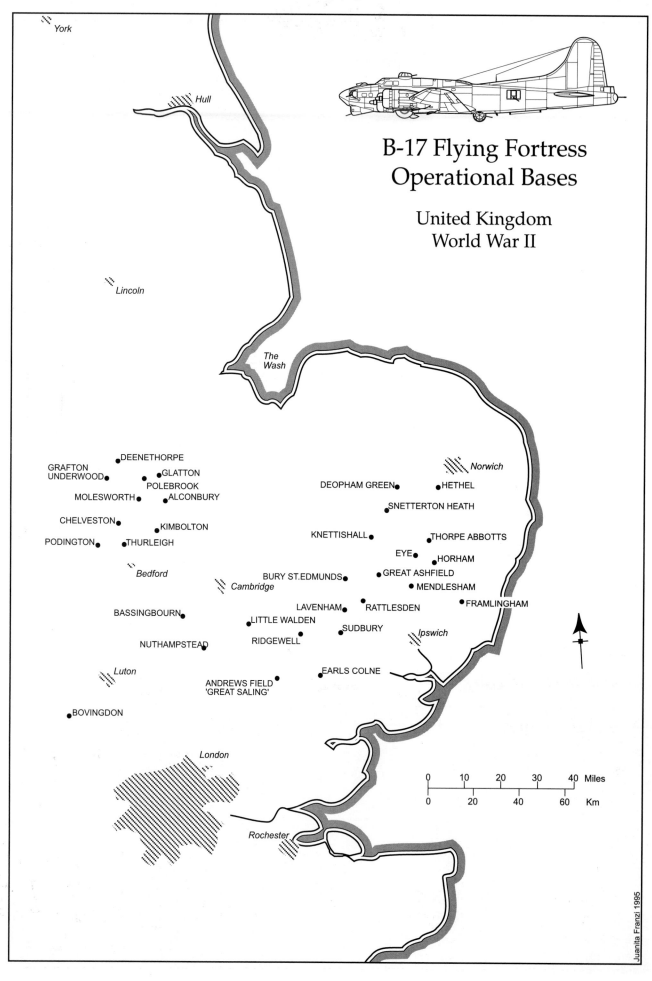

B-17 Flying Fortress Operational Bases

United Kingdom World War II

York

Hull

Lincoln

The Wash

DEENETHORPE
GRAFTON UNDERWOOD
GLATTON
POLEBROOK
MOLESWORTH
ALCONBURY
CHELVESTON
KIMBOLTON
PODINGTON
THURLEIGH
Bedford
Cambridge

Norwich
DEOPHAM GREEN
HETHEL
SNETTERTON HEATH
KNETTISHALL
THORPE ABBOTTS
EYE
HORHAM
BURY ST.EDMUNDS
GREAT ASHFIELD
MENDLESHAM
LAVENHAM
RATTLESDEN
FRAMLINGHAM

BASSINGBOURN
LITTLE WALDEN
SUDBURY
Ipswich
NUTHAMPSTEAD
RIDGEWELL
Luton
EARLS COLNE
ANDREWS FIELD 'GREAT SALING'
BOVINGDON

London

Rochester

| 0 | 10 | 20 | 30 | 40 Miles |
| 0 | 20 | 40 | 60 | Km |

Juanita Franzi 1995

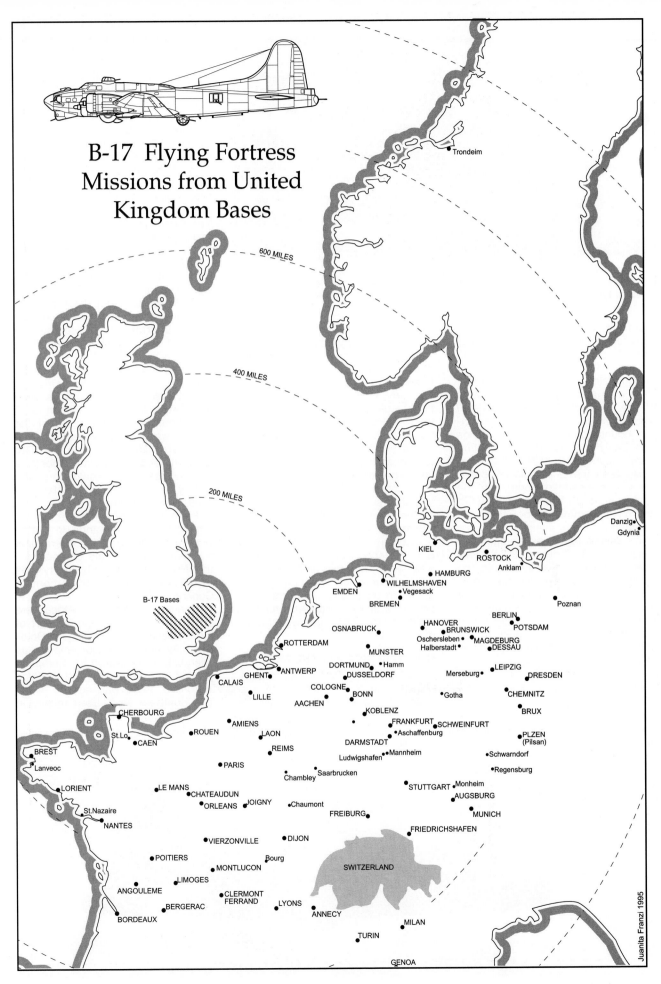

B-17 Flying Fortress
Missions from United
Kingdom Bases

600 MILES

400 MILES

200 MILES

B-17 Bases

Trondeim

Danzig
Gdynia

KIEL

ROSTOCK
Anklam

HAMBURG
WILHELMSHAVEN

EMDEN   Vegesack
                BREMEN

                                    Poznan

OSNABRUCK   HANOVER   BERLIN
              BRUNSWICK   POTSDAM
          Oschersleben   MAGDEBURG
MUNSTER   Halberstadt   DESSAU

ROTTERDAM

DORTMUND   Hamm   LEIPZIG
ANTWERP   DUSSELDORF   Merseburg   DRESDEN
GHENT
CALAIS   COLOGNE   CHEMNITZ
          BONN   Gotha
LILLE   AACHEN   BRUX

CHERBOURG   KOBLENZ
              FRANKFURT   SCHWEINFURT   PLZEN
St.Lo   AMIENS   Aschaffenburg   (Pilsan)
BREST   ROUEN   LAON   DARMSTADT   Schwarndorf
Lanveoc   CAEN   REIMS   Ludwigshafen Mannheim   Regensburg
                            Chambley Saarbrucken
LORIENT   PARIS   STUTTGART Monheim
          LE MANS   AUGSBURG
St.Nazaire   CHATEAUDUN JOIGNY   Chaumont
          ORLEANS   MUNICH
NANTES   FREIBURG
                    VIERZONVILLE   DIJON   FRIEDRICHSHAFEN
          POITIERS   Bourg
                    MONTLUCON   SWITZERLAND
ANGOULEME   LIMOGES
          CLERMONT
BERGERAC   FERRAND   LYONS
BORDEAUX   ANNECY   MILAN
                    TURIN
                GENOA

Juanita Franzi 1995

The B-17E bore the brunt of the USAAF's bombing activities in the Pacific until the B-24 Liberator began to take over. This B-17E is photographed over Gizo in the Solomon Islands in March 1942 at a time when the Japanese advance seemed unstoppable.

more significant were its contributions to the campaign against Rabaul from August 1942 and supporting American landings, particularly that on Guadalcanal in the same month. They were also involved in the Battle of the Bismark Sea (March 1943) when a major Japanese military convoy bound for the Solomon Islands was badly mauled, thus preventing Japanese efforts to hold the area.

By 1943 the B-17 was no longer the major American bomber in the Pacific, although two Groups (the 5th and 11th) continued operating until September.

It's difficult to assess the effectiveness of the B-17 during its relatively brief period of active service in the Pacific. It was certainly not enormously successful, although part of this can be blamed on the manner in which it usually had to go into action – in small numbers – thanks to the effects of relatively modest quantities of aircraft being available and a high level of unserviceability which more often than not reached a level of 50 per cent.

The elements also played their part with wildly changing conditions playing havoc with operational effectiveness. Heat, rain, dust, mud and ferocious tropical storms all hindered the aircraft which were operating from jungle strips in New Guinea and the New Hebrides with a minimum of maintenance facilities available. These conditions applied to all aircraft operating in the area during the first two years of the Pacific war, but the complex B-17 was more susceptible than smaller types.

The effect of weather on losses is interesting. During seven months of operations, the 11th BG lost six B-17s to enemy action and 12 to the weather. On one raid all three aircraft involved were lost when they were forced to ditch when they could not be successfully navigated through a nasty weather front, got lost and ran out of fuel.

NORTH AFRICA and THE MED

USAAF B-17s had a strong presence in the North Africa/Mediterranean Theatre of Operations (MTO),

operating under the control of the 9th Air Forces initially and from November 1943, the 15th Air Force, formed from parts of the 9th and 12th AFs and coinciding with their move from North Africa to Italy. The 15th Air Force had 21 Bombardment Groups on strength, of which six were equipped with B-17s and the remainder with B-24 Liberators.

The B-17 Groups were the 2nd, 97th, 99th, 301st, 463rd and 483rd, each comprising four squadrons and allocated to the control of the 15th Air Force's 5th Bombardment Wing. The 97th and 301st had been pioneering units with the 8th Air Force in England, but were transferred to the 15th AF in the second half of 1942.

Up until the end of 1943 there were more B-17s in 15th AF service than B-24s but from there the balance swung in favour of the Consolidated aircraft as it became the preferred heavy bomber for the Mediterranean theatre, while the B-17 enjoyed numerical superiority with the 8th Air Force operating from Britain.

A couple of famous names were associated with North African B-17 operations. The commander of the 12th AF was General James E Doolittle of Tokyo raid fame, while the 97th BG's CO was Major Paul Tibbets, a man whose name would enter the history books three years later in a B-29 over Hiroshima.

From February 1943 the Anglo-American air commands in northwest Africa merged for operational purposes into a single organisation, the Northwest African Air Forces (NAAF) commanded by American General Carl Spaatz.

Operations from North Africa were conducted from Egypt in support of the Operation Torch landings in October 1942 (the first major amphibious operation of the war in the European theatre) and the subsequent landings in Tunisia, Sicily, Salerno and Anzio. The bombers were used mainly in

The B-17's ability to get home after sustaining severe damage was legendary. This 97th BG B-17F somehow managed it after being rammed by a Bf 109 fighter and very nearly sliced in two. (Boeing)

tactical roles and were largely responsible for substantial interruptions to German communications in the area. A notable raid carried out by B-17s was recorded in March 1943 when aircraft from the 301st BG destroyed 30 acres of docklands at Palermo.

These operations were part of the overall effort which saw the surrender of the Italian Government in September 1943 and the subsequent battles in that country with German forces which effectively ended in early June 1944 with the fall of Rome. The taking of Italy gave the 15th AF new bases from which to operate, bases from which it was relatively easy to attack targets in southern Germany, Austria and the Balkans.

Operations against German targets in Italy meanwhile continued until into 1945 against lessening German resistance, although they had peaked between March and May 1944 during Operation Strangle, a co-ordinated air offensive against the railways, roads and shipping which were supplying the German forces.

The 15th Air Force's B-17s and B-24s operated in concert with the 8th AF bomber fleets based in England, sometimes sharing targets with the 'Mighty Eighth' and making an increasingly significant contribution to the Allies' bombing campaign against Germany as the war progressed.

THE MIGHTY EIGHTH

'The Mighty Eighth' – the nickname popularly applied to the US-AAF's 8th Air Force in Britain during World War II properly sums up the sheer size and strength of this massive fleet of bombers and fighters established with the single purpose of bombing Germany into submission.

At its peak in early 1945 the 8th Air Force was made up of 13 Bombardment Wings each comprising three or four Bombardment Groups in turn consisting of usually four squadrons. Of the 47 Groups which operated under the 8th's umbrella between 1942 and 1945, 29 were equipped with B-17 Fortresses and most of the remainder flew B-24 Liberators. Some Groups had both types on strength, 14 flew the B-24 exclusively while a handful of others were equipped with Martin B-26 Marauder medium bombers.

Supporting the bombers were three Fighter Wings of six Fighter Groups, each of these typically comprising four squadrons. The fighter units' equipment consisted mainly of Lockheed P-38 Lightnings, Republic P-47 Thunderbolts and North American P-51 Mustangs, the latter becoming the dominant type from 1944 due to its long range, a capability which allowed it to escort the bombers all the way to Berlin and back.

The classic plan of the B-17 photographed from another Fortress immediately above. (Boeing)

One that didn't make it, its wing folding upwards as the fuel tanks explode over Nis in Yugoslavia. The B-17 belongs to the 15th Air Force's 483rd Bombardment Group.

USAAF 8th AIR FORCE
B-17 BOMBARDMENT GROUPS GREAT BRITAIN 1942-45

Group	Squadrons	Wing	Date	Bases/Remarks
34th	4 7 18 391	93rd	11/42	Mendlesham; also operated B-24
91st	322 323 324 401	1st	11/42	Kimbolton, Bassingbourn
92nd	325 326 327 407	40th	09/42	Bovington, Alconbury, Podington
94th	331 332 333 410	4th	05/43	Bassingbourne, Earls Colne, Bury St Edmunds
95th	334 335 336 412	13th	05/43	Alconbury, Framlingham, Horham
96th	337 338 339 413	45th	05/43	Grafton Underwood, Andrews Field (Great Saling), Snetterton Heath
97th	340 341 342 414		08/42	Polebrook, Aug-Oct 1942 only
100th	349 350 351 418	13th	06/43	Podington, Thorpe Abbotts
301st	32 352 353 419		09/42	Sep-Nov 1942 only
303rd	358 359 360 427	41st	11/42	Molesworth
305th	364 365 366 422	40th	11/42	Grafton Underwood, Chelveston
306th	367 368 369 423	40th	10/42	Thurleigh
351st	508 509 510 511	94th	05/43	Polebrook
379th	524 525 526 527	41st	05/43	Kimbolton
381st	532 533 534 535	1st	06/43	Ridgewell
384th	544 545 546 547	41st	06/43	Grafton Underwood
385th	548 549 550 551	93rd	07/43	Great Ashfield
388th	560 561 562 563	45th	07/43	Knettishall
390th	568 569 570 571	13th	08/43	Framlingham
398th	600 601 602 603	1st	05/44	Nuthampstead
401st	612 613 614 615	94th	11/43	Deenethorpe
447th	708 709 710 711	4th	12/43	Rattlesden
452nd	728 729 730 731	45th	02/44	Deopham Green
457th	748 749 750 751	94th	02/44	Glatton
482nd	812 813 814		09/43	radio/radar tasks, also operated B-24
486th	832 833 834 835	4th	05/44	Sudbury, also operated B-24
487th	838 837 838 839	4th	05/44	Lavenham, also operated B-24
490th	848 849 850 851	93rd	05/44	Eye
493rd	860 861 862 863	93rd	06/43	Debach, Little Waldon; also B-24s

Notes: The table shows (left to right) the Bombardment Group, the squadrons operating within it, the Bombardment Wing which controlled it, the date of its establishment in the UK, and the bases from which it operated.

Other B-17 units based in UK were: 5th Emergency Rescue Squadron (from March 1944); 15th Photographic Squadron; 422nd Bombardment Squadron (later 858th and 406th BS), night leaflet drops 1943-44; 652nd BS (reconnaissance) from late 1944; 803rd (later 36th) BS, electronic countermeasures from mid 1944.

Each Group had a nominal 72 aircraft on strength, simple arithmetic telling us that the 8th Air Force could have more than 4,000 aircraft in England at any one time once all the operational units were established. Of these, about half were B-17s.

The logistics supporting the 8th Air Force are impressive. The 8th occupied 68 airfields on England's eastern side in the counties of Cambridgeshire, Huntingdonshire, Essex, Hertfordshire, Bedfordshire, Northamptonshire, Norfolk, Suffolk and Lincolnshire. Most of these airfields had to be built, their construction resulting in one of the biggest civil engineering projects ever undertaken in Britain. When the programme was at its peak in 1942, construction of a new airfield was being started every three days.

Each one cost about £500,000 and the quantities of concrete, tarmac, water and sewer pipes, electrical conduit, bricks and other building materials used was colossal. Add the necessary maintenance, administration, operational and accommodation buildings to the basic requirements of runways, dispersals, perimeter tracks and access roads and the size of the undertaking can just about be imagined. The fact it was all done in an extremely short time adds to the achievement.

Bombing Theory

With the entry of the United States into the war in December 1941, plans which had already been discussed were immediately put into place to mount a strategic bombing campaign against Germany by the USAAF with the bombers operating from Britain. This would be part of a two pronged attack – the Americans for the first time putting their beloved daylight, high altitude precision bombing theories into practice; and the RAF bombing Germany by night. Both considered that this intensive bombing alone would be sufficient to bring Germany to its knees, removing the need for a costly amphibious invasion and then an equally costly advance across Europe by Allied troops.

This turned out to be a false hope, and although the strategic bombing of Germany was a major contributor to the eventual victory, it was not able to do the job alone. The soldier on the ground would still be needed even if the enemy's ability to manufacture goods or supply fuel to its war machine had been substantially reduced by the bombers.

The American bombing campaign actually took some considerable time to get into its stride and then after substantial losses had been inflicted on the bombers by *Luftwaffe* fighters. The theory of mutual self defence being provided by heavily armed bomber formations did not live up to expectation and it became clear that the use of escorting fighters was the best defence.

It wasn't until well into the second half of 1944 when German fighter resistance began to wane, when long range P-51 Mustangs were available to fight off the *Luftwaffe* throughout even the longer missions, and when enormous numbers of bombers were available that the idea of daylight precision bombing began to look viable. As for the theory that bombing would remove the need for invasion, the fact that more than two-thirds of the tonnage dropped by the 8th Air Force was delivered *after* the D-Day landings in June 1944 speaks for itself.

The British had discovered the perils of unescorted daylight bombing long before the 8th AF arrived in Britain and had switched to night 'area' raids in early 1942. For the USAAF there was more at stake than immediately met the eye. Political considerations entered the equation as well, a major influence being the desire of its hierarchy that the Army Air Force should become an entirely independent entity. If the theory of daylight precision bombing could be proved viable, that independence might be more easily gained. The same background applied to B-29 Superfortress operations against Japan from 1944. As is discussed in the B-29 section of this book, success was difficult to come by in that area of operations as well, but the Air Force won its independence regardless, in 1947.

Defensive Formations

The basis of the daylight bombing theory was the 'mutual defence' formation of the bombers, where defensive firepower could be concentrated on any particular area of the sky around the bombers. Several variations on these formations were used over the years as it was discovered what worked and what didn't as the number of aircraft per squadron and group increased.

The earliest formations were based on six aircraft flying in two staggered, inverted vics. This changed in September 1942 with the introduction of 18 aircraft Group formations, comprising two squadrons of nine aircraft. The squadrons flew in three unstaggered vics with the lead squadron 500 feet below the second which was slightly behind and echeloned towards the sun. This formation tended to close down fields of fire and thus reduce mutual support and also made it very difficult for the outer aircraft to keep up with the remainder of the formation when turning.

A revised 18 aircraft formation was introduced in December 1942. Split into squadrons of six, the aircraft were stacked towards the sun with the leading squadron in the centre and the high and low squadrons behind. Each group of 18 aircraft flew slightly above the one in front echeloned towards the sun.

Heavy losses to German fighters caused another change in March 1943 and the introduction of the 54 aircraft combat wing formation comprising three groups of 18 aircraft. The lead formation flew in the centre with the two trailing formations above and below with the aircraft spread out over a width of 2 kilometres (1.25 miles), a depth of 800 metres (half a mile) and a length of about 600 metres. This formation provided much better firepower and mutual support between the aircraft but was unwieldy with aircraft in the high group finding it difficult to keep the leader in view.

The next major variation of 8th AF formations was introduced in early 1944 and resulted from the arrival of large numbers of escort fighters and a reduction in the need for concentrated defensive firepower as a result. Comprising three groups of 12 aircraft, the formation was lead by the middle group with upper and lower groups trailing slightly behind, the effect being one of an echeloned vic. The lead aircraft was usually equipped with radar bombing equipment, although this became more readily available as 1944 progressed and was fitted to many B-17s.

(above) 381st Bomb Group B-17Gs formate over England. Note the accompanying P-51B Mustang fighter at the top of the picture. The ability of the Mustang to escort the bombers deep into Germany and back greatly increased their chances of survival from 1944.

(right) This 463rd BG Fortress somehow managed to fly on for a short time after its whole nose had been blown off by flak over the target.

Curtiss LeMay developed the concept of 'pattern bombing' by which a formation's bombs were dropped simultaneously on their leader's signal. The idea was to cut out inaccuracies in bomb aiming from individual aircraft, depending instead on the experienced leader's accuracy.

(below) Patterns in the sky. B-17Fs of the 390th BG make their way across Europe accompanied by the vapour trails created by their escorting fighters.

A final revision to the formation was made in February 1945 as the B-17s encountered very heavy anti aircraft fire over Germany. The idea was to have a formation which made life more difficult for the German gunners on the ground and which was also easier for the escorting fighters to accompany. This 36 aircraft group formation comprised four squadrons of nine aircraft staggered nearly directly above each other and resulting in greater depth but reduced width and length, thus giving the anti aircraft gunners less to aim at. This formation was also easier to hold than the previous ones.

The theory of mutual defensive firepower depended to a very large extent on formation discipline, but holding them intact was a very difficult exercise for the bomber pilots, requiring intense concentration and substantial physical effort. With their aircraft being buffeted by not only natural turbulence but also by the turbulence created by the airframes and propellers of the many bombers around him, the pilots tired quickly.

The formations therefore lost their 'tightness' more often than not, increasing the risk of collision and reducing the effectiveness of the combined defensive fire. A straggler was easy meat for the *Luftwaffe*, bearing in mind that for the first 18 months of the 8th AF's campaign there was little or no protection by escorting fighters for the greater part of the mission including while the bombers were near their targets.

Help didn't arrive until early 1945 with the fitting of the 'Formation Stick', an electrically operated power boost for the control column operated by a pistol grip device attached to the column. This eased the pilot's workload considerably but it hadn't been available when it was really needed, two or so years earlier.

The Influence of Curtiss LeMay

The first of the revised combat formations mentioned above came about at the instigation of Col Curtiss LeMay, in late 1942 the commander of the 305th BG at Grafton Underwood in Northamptonshire. LeMay's influence would be widely felt in the 8th Air Force as he was responsible for many of the changes and innovations which were introduced as the campaign rolled on.

Apart from his work on combat formations, one of LeMay's other major contributions was the introduction of the 'pattern bombing' technique, in which the aircraft's pilots had to resist the overwhelming urge to take evasive action when flak or fighters came up during the bombing run. The aircraft had to remain flying

When the bombers created these trails they acted like a beacon to German fighters, a deadly situation if the B-17s had no escort. In this case escorting fighters are evident, criss-crossing the bomber formations.

straight and level and hold their positions relative to the leading bomber, releasing their bombs simultaneously when a smoke marker was dropped from the lead aircraft.

This negated individual sighting errors by bomb aimers in separate aircraft and meant that providing the lead aircraft's calculations were accurate, all the bombs from a formation would fall on and around the target. With more experienced crews in the lead aircraft, accuracy should theoretically be good, although if an error was made, it meant the whole formation's bombs missed their mark. LeMay first tried the technique with his own group (the members of which thought they were about to commit suicide!) but it was successful and was soon adopted by the other units.

LeMay was highly regarded by both his superiors and those who served under him and he went on to greater things with the B-29 Superfortress units which flew against Japan later in the war.

B-17s to Britain

Agreement to introduce a large USAAF presence to Britain was

reached in January 1942 at a meeting in London. The 8th Air Force was formed in the USA shortly afterwards and began training at its home base in Florida. British headquarters were established at High Wycombe in Buckinghamshire with Gen Ira Eaker in command of bomber operations. Maj Gen Carl Spaatz was overall commander of the 8th Air Force in his capacity as Commander of American Air Forces in Europe.

The first unit selected to deploy to Britain was the 97th Bomb Group, then equipped with B-17Es. The order to move was issued in mid May 1942 and after a delay caused by the necessity to temporarily divert to the Aleutian Islands following a Japanese attack, the 97th's aircraft finally arrived at their Polebrook, Northamptonshire, base in July 1942.

The 97th remained in Britain only until October of the same year but in the meantime conducted the 8th AF's first attack on 17 August when 12 B-17Es and Fs struck the marshalling yards at Rouen in north-western France (the place, incidentally, where Joan of Arc had been burnt at the stake in 1431). The raid was led by

B-17Gs 42-31718 and 42-31447 of the 337th and 338th Bomb Squadrons, 96th BG.

General Eaker in B-17E 41-9023, the regular aircraft of the group's commander, Col Frank Armstrong. Armstrong took over the B-17E of Major Paul Tibbets, who flew as copilot to his CO.

Flown from Grafton Underwood (a satellite airfield to Polebrook) the raid was regarded as successful as some damage was inflicted on the target and all of the B-17s returned safely. A diversionary sortie by six of the 97th's Fortresses successfully drew German defences away from Rouen.

The buildup of the 8th AF in Britain continued during 1942, slowly at first due to the enormous logistics involved in building airfields and establishing the substantial infrastructure required. By the end of the year only seven bomber groups were in place but that number grew rapidly throughout the following year to the point that by December 1943 the number had increased to 26. Another 14 were established in the first few months of 1944.

One point of contention about US bomber operations from Britain has been American claims of a lack of co-operation from RAF Bomber Command and its Commander-in-Chief, Air Chief Marshall Arthur Harris. The British had no time for daylight preci-

sion bombing from high altitudes and had already switched to night operations. They tried to convince the Americans to do the same, but the doctrine of daylight operations was well entrenched in the American bomber bosses' psyche – it was their Holy Grail, and no amount of persuasion would move them from it.

The day/night philosophies of the two air forces should have set the scene for a co-ordinated 'around the clock' effort, complementing each other and straining German resources to the limit. Even though this in effect happened, it wasn't as a result of any large amount of co-operative

effort. There was little co-ordination between the USAAF and RAF bombers, Arthur Harris preferring to go his own way, refusing to bomb certain targets which were then given to the Americans.

To The Fatherland and Problems

The 8th AF's 1942 combat missions were restricted to bombing targets in Nazi held territory in western Europe (including many against U-boat bases on the French coast) but not in Germany itself, mainly due to the relatively meagre number of aircraft which were available until large numbers began to arrive from mid 1943. Germany itself was finally attacked on 27 January 1943 when 64 B-17s led by the 306th BG's Col Frank Armstrong attacked the naval base at Wilhelmshaven on the north coast.

Fifty-eight aircraft reached the target and only one B-17 was lost, despite the close attentions of unhealthy numbers of German fighters. American gunners claimed 22 enemy fighters shot down, but the figure was in fact seven. The overclaiming of 'kills' by gunners would continue to be a problem and usually resulted not from any deliberate exaggeration but from the fact that at a given time the gunners in several bombers could all be firing at one particular fighter. When they saw it go down, they tended to assume they were the one responsible for its demise, so the kill was claimed several times over.

This first up success was something of a false dawn because soon the German fighters would be hacking the 8th AF's B-17s and B-24 Liberators out of the sky. Another raid into Germany on the Focke-Wulf factory at Bremen in April resulted in the loss of 16 B-17s, one to flak and the remainder to fighters. A further 48 Fortresses were damaged out of a total force of 117 aircraft.

At this time, the 8th Air Force was preparing to unleash itself on Germany, but its effectiveness would be questioned mainly due to the high level of losses sustained during the

An interesting 8th Air Force B-17E, 41-9112 'The Dreamboat', used as an armament test 'mule'. Note the B-24 Liberator tail turret mounted in the Fortress's nose! 'Dreamboat' wasn't flown on operations.

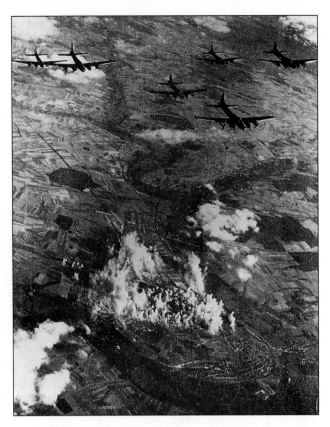

Bombs away! An example of pattern bombing by B-17Gs of the 303rd BG.

8th Air Force B-17s give the Luftwaffe fighter base at Amiens-Glisy, France, a thorough going over in August 1943.

course of 1943, an overall total of 967 from 22,800 sorties. This represented a loss rate of 4.24 per cent; by comparison RAF Bomber Command recorded a loss rate of 3.6 per cent during the same period.

Some examples of the 8th AF's more extreme losses during 1943 are 22 out of 69 lost over Kiel in June, and in the last week of July ('Blitz Week') 100 B-17s out of a total operational force of 330 were either lost on operations or had to be written off because they were so badly damaged. In September 1943, 50 out of 388 bombers were lost on a raid on Stuttgart, and to rub it in, hardly any of them managed to find the target.

Overall losses to the American bombers in the second half of 1943 were running at the unacceptable rate of 10 per cent per mission, with the lowest and therefore most exposed squadron position in the group formation often being wiped out completely. This position came to be known as 'Purple Heart Corner' by the crews.

Schweinfurt

The 'twin' raids carried out on Schweinfurt and Regensburg in August 1943 and the follow up attack on Schweinfurt two months later well illustrate the problems facing the 8th Air Force. The first raid cost 60 out of 363 bombers with more written off afterwards while the second was

even more expensive when 77 were lost and a further 133 damaged out of a total force of 291 aircraft.

Operations were temporarily halted after the second Schweinfurt raid. Overall losses had been three times greater than expected, this causing not only the obvious problems associated with such a large loss of equipment and men, but also those associated with morale. A break was needed to reassess the situation and to catch a few breaths.

The Schweinfurt and Regensburg raid of 17 August 1943 was one of the most important of the war and was a major test of both the American bombing philosophy and the German defences. Regensburg was a major centre of Messerschmitt Bf 109 fighter production and factories at Schweinfurt produced ball bearings, a seemingly insignificant item but without which all manner of equipment could not operate. To destroy these centres would put Germany's war production into chaos.

The raids would be conducted over relatively long ranges, with Schweinfurt more than 400 miles (640km) from the B-17s' bases and Regensburg another 100 miles (160km) further away. The idea behind having one large force heading out and then dividing into separate forces for each target was to split the enemy's defences. The plan was for the Regensburg formation of 146 bombers from

the 4th Bombardment Wing to depart England first, with Schweinfurt's 217 aircraft of the 1st BW following close behind. The Regensburg force was led by Curtis LeMay. To further confuse the enemy fighters, the Regensburg force would not return to England after dropping its bombs, but turn south and fly over the Alps and the Mediterranean to airfields in North Africa.

The length of the mission necessitated an early morning start, and on the morning in question the eastern part of Britain was covered in fog and mist. After some deliberation as to whether or not call off the mission, the go ahead was given and the 4th BW's B-17s departed more or less on time. At this point the whole plan came unstuck as the 1st Wing did not get off the ground until some four hours later due to the fog and mist.

The delay defeated the whole purpose of the plan and threw the complex escort fighter arrangements which had been made into disarray. As it was, the accompanying P-47 Thunderbolts could only provide an escort as far as the German border due to fuel limitations, at which point the bombers still had 300 miles (480km) to travel to Regensburg.

The splitting of the force gave the German fighters plenty of time to land, replenish, refuel and rise again to meet the second wave. The losses to both bomber wings were substan-

(above) B-17s of the 305th BG over Schweinfurt during the second raid on 14 October 1943.

(below) The ball bearing plants at Schweinfurt after the second raid on the facility in October 1943. The damage inflicted was greater than the first raid two months earlier but B-17 losses were still substantial – 77 out of 291.

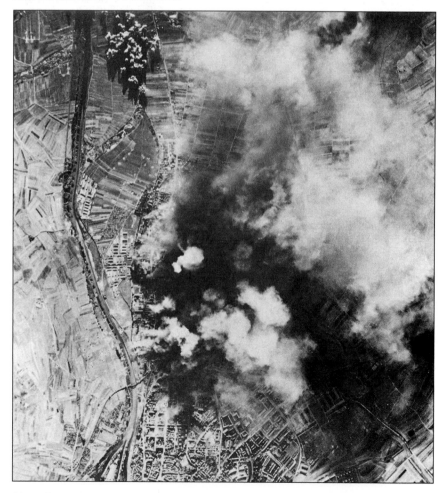

tial as not only did the *Luftwaffe* have early warning of the approaching bombers thanks to radar, they also had plenty of time to attack as the aircraft streamed across Europe on their long flight. Sixty bombers were lost in return for 21 fighters claimed shot down.

The 4th BW managed to get 122 bombers through to Regensburg dropping 250 tonnes on the Messerschmitt factory but with only a temporary effect on production. Few of the plant's vitally important machine tools were damaged but the effect was the loss of production of about 800 fighters before full scale manufacture could resume. An unknown bonus was the destruction of fuselage jigs for the Me 262 jet fighter, at that time something the Allies had no idea existed.

The 1st BW's attack on Schweinfurt resulted in 80 hits on two of the three main ball bearing plants, temporarily reducing production by half. It would have caused some dismay to the American generals had they known that production actually increased the following month.

The Schweinfurt/Regensburg raid resulted in an overall loss rate of 16 per cent, a clearly unacceptable figure and one which the 8th AF could not sustain for long.

Despite this, a second raid on Schweinfurt was carried out on 14 October 1943. This time 291 B-17s took part and no fewer than 77 – or 26 per cent – failed to return and most others were damaged from the onslaught provided by the German fighters including twin engined Me 110s carrying rockets and attacking from astern. As before, the escorting P-47s could travel only as far as the German border.

Most of the formations were broken up and many B-17s were shot down before they got anywhere near the target. The fighter attacks eased off as the bombers neared Schweinfurt and a virtually unopposed drop was able to take place, resulting in a substantial amount of damage being inflicted on the ball bearing plants. As the B-17s turned for home, the German fighters renewed their attacks, inflicting more serious damage on the bombers, causing them to struggle home as best they could.

Fighter protection could not even be provided for the last leg of the return journey as bad weather left them grounded. The American gunners claimed they had shot down 288 German fighters during the course of the raid. This was later reduced to 186 but total *Luftwaffe* losses for that day were later revealed to be only 38 with a further 20 aircraft damaged.

The second Schweinfurt raid had a

great effect on both the attackers and the defenders. For the Germans, the damage caused by the bombers resulted in the decentralisation of the ball bearing and other industries; for the Americans it time to pause and think again about the whole philosophy of their bombing campaign.

From then on, unescorted missions deep into Germany were avoided. A degree of salvation appeared in the form of the P-51 Mustang in December 1943 and with it was applied a new tactic in which the fighters would not only protect the bombers by accompanying them, they would fly ahead of the B-17s and B-24s to actively seek out German fighters and engage them in combat. This was in itself turning point, and from mid 1944, the 8th Air Force's fortunes began to change for the better.

The Big Week

A major campaign against the German aircraft industry was mounted in February 1944. Aircraft production had been steadily increasing during 1943 despite Allied attempts to slow it down. The February raids also provided one of the few examples of co-ordinated efforts by the British and American air forces.

Now commanded by James Doolittle, the 8th AF sent out more than 1,000 bombers and several hundred fighters against the German aircraft factories during what became known as the 'Big Week'. Between February 20 and 25, the 8th flew 3,300 sorties against a dozen factories, supported by 500 sorties by the bombers of the 15th Air Force based in Italy, plus more than 3,500 fighter sorties. The RAF contributed five massive night attacks to the campaign.

The Big Week cost the Allies a loss rate of six per cent – just about acceptable especially by comparison with recent results – and in return set German aircraft production back for a period of two months at a critical time. Despite this, fighter production quickly recovered and reached a peak in the final months of 1944. By then it had in fact more than doubled since mid 1943! The continued appearance of the *Luftwaffe* in substantial numbers caused some consternation to the Allies, whose intelligence had wrongly suggested that aircraft production had been severely disrupted by the bombing raids. Efforts to engage the enemy in air-to-air combat were appropriately

The danger didn't always come from enemy fighters or flak. This 94th BG B-17 is over Berlin and has had its port tailplane knocked off by a bomb falling from another Fortress. Another bomb can be seen falling perilously close to the damaged B-17.

(above) A 384th BG B-17G drops its load over Nuremburg in February 1945.

B-17F 42-30338 'Cabin In The Sky' of the 390th BG's 571st Squadron.

Berlin for the first time on 3 March 1944 was deeply symbolic to both sides of the war. For the Americans, it meant its bombers could go just about anywhere in the Third Reich with fighter protection; and for those Germans prepared to face the truth it meant the beginning of the end.

Three days after that brief appearance, the first 8th AF raid on the 'Big City' in large numbers took place. The small force of 29 B-17s which had attacked Berlin on the 3rd were part of a larger force intending to bomb the German capital on that day but the mission had been scrubbed due to poor weather and the aircraft recalled whilst *en route*. The 29 B-17s which carried on had not heard the recall order. Back in England, the hierarchy realised what had happened and left some of the escorting Mustangs with the bombers. Five B-17s were lost, but the propaganda value of that 'accidental' first mission to Berlin was high.

stepped up and the Americans claimed the destruction of no fewer than 800 German fighters in February and March 1944.

Regardless of how many Bf 109s or Fw 190s could be manufactured to replace these losses, the German fighter forces found themselves short of fuel over the last six months of the

European war (thanks largely to the USAAF and RAF bombers) and were also very short of the most precious commodity of all – pilots, especially experienced ones.

B-17 To Berlin

The appearance of a small number of B-17s and P-51 Mustangs over

An interesting innovation which appeared in 1944 was the GB-1 guided glider bomb, illustrated here under a test B-17. Basically a 2,000lb bomb with wings, booms, tailplane assembly and radio receiver bolted on, the GB-1 was first used on 28 May 1944 against the Cologne Eifeltor marshalling yards. One hundred and nine were launched but accuracy was poor. A further 1,000 GB-1s were subsequently launched against other targets but a lack of accuracy remained a major problem. (Boeing)

A US Navy PB-1 Fortress used for experimental purposes, in this case carrying a quarter scale model of a Grumman Bearcat fighter under its fuselage.

The real USAAF assault on Berlin began on 6 March when 700 bombers (B-17s and B-24s) escorted by 800 fighters raided the city. A huge battle developed in the air as German fighters attacked the massive bomber stream as it made its way across Germany. Despite the presence of escort fighters, 23 B-17s were lost well before the target had been reached with waves of Bf 109s, Bf 110s and Fw 190s intercepting the bombers and the Mustangs fighting to keep them away. Even as the leading group of B-17s were on their bombing runs, the opposing fighter forces were hard at it in the skies above Berlin. Five bombers were shot down during this phase of the battle, a battle which continued for some time as the 100 miles (160km) long bomber stream poured over the city.

The raid resulted in the loss of 69 out of the 702 bombers which had reached Germany (776 had originally set out from Britain), a rate of 9.8 per cent. A further three made it home but were subsequently written off and 102 others were badly damaged. The primary objective of the mission had been to destroy some factories in the southern part of Berlin. Due to cloud cover this part of the mission failed and bombs were dropped mainly on targets of opportunity, but the secondary aim – to destroy as many German aircraft as possible – was successful, with more than 40 being shot down.

This raid was followed up with another two days later when 600 more American bombers attacked Berlin. Once again, the shooting down of *Luftwaffe* aircraft was at least as important as the bombing itself.

The End

Eighth Air Force bombers continued to pound strategic targets in Germany throughout the remainder of the European war with the enemy oil industry coming in for special attention.

At the end of it all on 7 May 1945, the Mighty Eighth had logged some impressive statistics – well over 300,000 sorties flown and 700,000 tonnes of bombs dropped. On the other side of the coin was the loss of nearly 4,400 bombers of all types in action plus another 1,000 or so from other causes. The big question as far as the USAAF was concerned is whether or not the concept of high altitude daylight precision bombing was proven to be a sound one. It certainly had its difficulties, but at the end of the day the Germans were beaten, largely thanks to the efforts of the crews flying the B-17s and other bombers which eventually dominated the skies over Europe.

POSTWAR FORTRESSES

Use of the B-17 did not stop with the end of World War II. Postwar

Israel acquired three B-17Gs which operated in the 1948-49 war prior to the formation of the state of Israel. One was used to bomb Cairo. (Boeing)

Saab converted eight B-17s to commercial transports shortly after the war for operation on trans-Atlantic and other services. The first Stockholm to New York service was flown in July 1945. One of them is pictured at New York's La Guardia Airport on the inaugural service. (SAS via Boeing)

A major postwar user of the B-17 was France's l'Institut Geographique National, which operated a fleet of aircraft for photographic survey duties for many years. This particular one – the former B-17G 44-85643 – was the last in service but met an unfortunate end when it crashed during filming of the movie 'Memphis Belle' in 1989.

Trans World Airlines operated this executive conversion B-17G (ex 44-85728) for short period from 1946. The conversion was performed by Boeing and the aircraft dubbed the Model 299AB. It was sold to the Shah of Persia in 1947. (Boeing)

USAF variants created by conversion and their roles are discussed in the previous chapter. The last operational USAF Fortress was Douglas built B-17G 44-83684 'Piccadilly Lily' (famous for its role in the TV series *12 O'Clock High*) which recorded its final flight in military service in August 1959 when it was flown to Davis-Monthan AFB in Arizona. This historic B-17 is now preserved at the Planes of Fame Museum at Chino, California.

Several South American nations flew B-17s after the war, notably Brazil (SB-17s), Chile (for search and rescue duties) Bolivia and Dominica, the latter pair keeping some aircraft in service well into the 1960s. In Europe, Fortresses flew in French, Danish, Portuguese and Dutch colours.

A significant postwar operator of the Fortress was the Israeli Air Force, which acquired three B-17Gs which operated in the 1948-49 war prior to the formation of the state of Israel. One of the Israeli B-17s was used to bomb Cairo.

Other Fortresses were used for more peaceful purposes. About 60 B-17s had made forced landings in neutral Sweden during World War II. More than 30 of these were returned to USAAF in mid 1945 but eight others were purchased from the US government for the symbolic sum of $US1.00 each and converted into airliners by Saab.

The conversion involved removing all gun turrets and gun ports, installing 14 passenger seats and incorporating windows into the fuselage sides. An elevator was installed for loading freight into the former bomb bay. The converted Fortresses were operated by the Swedish ABA airline in partnership with Denmark's DDL on trans-Atlantic services. The first Stockholm to New York (via Iceland and Canada) service was flown on 27 July 1945 and services were soon expanded to include Rio de Janeiro, Addis Ababa, Cairo, Moscow, Paris and Zurich.

France's *l'Institut Geographique*

National was a major user of B-17s on photographic mapping survey work for many years from 1947, a dozen aircraft operating with that organisation between then and the retirement of the last example in 1989. Most of the others had been withdrawn from service by the mid 1970s.

Other civil users of the Fortress included Trans World Airlines which had an executive conversion example (Boeing Model 299AB) used for commercial services from 1946; other examples were converted to executive transports for corporations in the USA; and still others were revamped as water bombers to fight forest fires. As recently as 1969, Aero Flight Inc in the USA converted one of these aircraft to Rolls-Royce Dart turboprop power. This aircraft (the former B-17F 42-6107) unfortunately only survived for a year in its new guise, crashing while undertaking a firebombing mission in Yellowstone Park, Wyoming, in October 1970.

Bombing of a different kind. B-17F N17W (ex 42-29782) on a tree spraying sortie over Lansing, Michigan in 1953. This Fortress survives today and spent many years operating as a fire tanker. It also appeared in several films including '1000 Plane Raid', 'Tora Tora Tora' and 'Memphis Belle'. (Boeing)

AVRO
LANCASTER

Starkly silhouetted against the evening light, Lancasters wait at dispersal for their next mission. The RAF's best heavy bomber of World War II completed 156,000 wartime sorties, dropping more than 608,600 tons of bombs. (via Neil Mackenzie)

AVRO LANCASTER

It would probably not be too far fetched to suggest that of all the bombers which took part in World War II, the Avro Lancaster was the most successful. Although best remembered for some of the famous actions in which it took part – the dams raid, the sinking of the *Tirpitz*, the destruction of the Beilefield Viaduct, its use by the Pathfinder Force and for numerous special missions using weapons such as the 12,000lb (5,443kg) *Tallboy* and extraordinary 22,000lb (9,980kg) *Grand Slam* 'earthquake' bombs – the Lancaster's real place in history comes from simply being very good at the role for which it was designed ... to carry large loads of bombs to its target routinely.

To achieve this, the Lancaster first had to be available, it's ease of manufacture, maintenance and repair contributing to this vital factor. It also had to have suitable performance in order to fulfil its role and this it did,

particularly in the area of payload/range where it bettered all its contemporaries including the equally famous Boeing B-17 Flying Fortress. It wasn't until that technological marvel of its time, the B-29 Superfortress, came along that the Lancaster was outshone.

The ability to carry useful payloads over practical distances was the Lancaster's greatest attribute. In 1942 and for some years after that a bomb load of 14,000lb (6,350kg) was considered massive, yet a Lancaster could carry that over a range of 1,660 miles (2,670km). By comparison, a Boeing B-17G Flying Fortress had an absolute and rarely carried maximum bomb load of 12,800lb (5,800kg) which could be carried just 1,100 miles (1,770km); its typical load on a mission over Germany was considerably less. With 7,000lb (3,175kg) of bombs aboard and an auxiliary tank fitted the Lancaster could travel 2,680 miles (4,310km).

Some statistics associated with the Lancaster's wartime career underline its value. The type took part in every major night bombing attack on Germany, one aircraft was lost for every 132 tons of bombs dropped (compared with 86 tons for the Handley Page Halifax and 41 for the Short Stirling) and Lancasters completed 156,000 wartime sorties dropping 608,612 tons of bombs in the process. No fewer than 61 Royal Air Force operational squadrons flew Lancasters between 1941 and 1945 along with numerous training and secondary units.

Apart from having the ability to successfully perform its intended role, the Lancaster should also be remembered for the part it played in the formative days of the black art of electronic warfare. With all aspects of radar and electronics developing as the war progressed, the Lancaster was the first application for many of the various radar, navigation, fighter

One of only two airworthy Lancasters in existence, the Battle of Britain Memorial Flight's PA474 'City of Lincoln' displays its majesty to another air show crowd. This photograph was taken at Lakenheath way back in June 1970, shortly after the aircraft had been restored to airworthiness. (Philip J Birtles)

Two of Avro's finest before the Lancaster, the 504 trainer and the Anson. Both were built in very large numbers.

detection, blind bombing and jamming devices we now take for granted in modern warfare. As such, it represented both the last of the old generation of bombers and the first of the new.

A Proud Heritage

Englishman Alliot Verdon Roe was one of the true pioneers of flight and is credited with making the first sustained flight of an all British aeroplane in 1909. Born in 1877, the son of a Manchester surgeon, Roe's early adult years were spent following a basically engineering and draughting course interspersed with time in the silver mining industry in British Columbia (Canada) and at sea. Like so many young men of that time, Roe became smitten with the flying bug and achieved his first success in 1906 when he won £75 in prizes for the flights of three model aircraft in a competition sponsored by the *Daily Mail* newspaper.

The step to full size aircraft was a logical one, his first design, the canard 'Roe I Biplane' making some tentative hops in June 1908. The redesigned Roe I Triplane followed in 1909, this aircraft achieving successful controlled flight and setting the scene for a large number of successful military and civil designs over the next half a century.

A V Roe & Co was formally established on 1 January 1910 with workshops in Manchester and flying at the famous Brooklands race track where the new company (the name of which was quickly contracted to simply 'Avro') also operated a flying school.

After building and flying a series of triplane and biplane designs, Avro's first taste of success came in 1912 with the Type 500 two seat training biplane. Impressed with its performance, the War Office ordered two and from the basic design was developed the famous Avro 504, first flown in 1913. The 504 became the RAF's standard trainer in World War I and went on to be continuously developed and built in several major production variants for the RAF and other users, the last example not coming off the production line until as late as 1932.

The aircraft's versatility allowed it to be fitted with ever more powerful engines – early models had an 80hp (60kW) Gnome rotary and the final version a 180hp (135kW) Lynx radial – and to be used for numerous roles other than trainer, roles which included shipboard operation, night fighting, anti Zeppelin patrolling, communications, as a floatplane and for joyriding and stunt flying by civilian operators. Some 10,500 504s were built by 23 manufacturers in Britain, Ireland, Scotland, Canada, Belgium, Denmark and even Japan.

If the 504 established Avro as a major designer and producer of aircraft, subsequent offerings built on that and included some of aviation's classics such as the Avian (as flown by Bert Hinkler from London to Darwin and other pioneer aviators), the Type 618 Ten (a licence built Fokker F.VIIb/3m airliner), the Anson multi purpose civil and military twin built continuously from 1935 to 1952 in numerous versions, the Lancaster, Lincoln, Shackleton and Vulcan delta winged jet bomber. The last entirely Avro design to be built was the successful Type 748 44 to 50 seat twin turboprop regional airliner which began life in 1960 as the Avro 748 and subsequently became the Hawker Siddeley and then British Aerospace 748 as the British aircraft industry was nationalised and rationalised over the years. The 748 finally went out of production at BAe's Manchester facility – a legacy of Avro's long term association with that city – in 1987.

A V Roe and Co became a Limited company in 1913 but it came under the control of the Armstrong Siddeley Development Co Ltd in 1928 (this organisation also controlled Sir W G Armstrong Whitworth Aircraft Ltd at Coventry) when Roe sold his shareholding. His ambition was to build flying boats – something he'd not been able to achieve at Avro – so with the money obtained by the sale of the company he formed he was able to take a controlling interest in S E Saunders Ltd, a company long specialising in marine craft and flying boats. The company's name was changed to Saunders-Roe Ltd the following year and went on to produce flying boats such as the Cloud, London, the postwar giant Princess, the SR.A/1 jet powered flying boat fighter, the SR.53 supersonic jet and rocket powered fighter and the Skeeter light helicopter.

Roe was Knighted in 1929 and remained President of Saunders-Roe until his death in 1958 at the age of 80.

Roe's departure from Avro allowed two dominant figures in the British industry to emerge. The first was Roy Dobson who had joined Avro as an engineer in 1914 and progressed to Works Manager in 1919, General Manager in 1934 and Managing Director in 1941. Dobson went on to become Chairman of the Hawker Siddeley Group, Avro's parent company from 1935.

An interestingly lit night engine runup by a Lancaster reveals some undercarriage, nose and underside detail.

Then there was Roy Chadwick, one of Roe's earliest employees (from 1911) and a man who quickly established himself in the design department of the company, soon as Chief Designer. Chadwick was responsible for many of Avro's best known designs (including the Anson, Lancaster, Lincoln and early Vulcan concepts) and remained the company's Chief Designer until his untimely death in the crash of the Tudor II transport prototype in 1947. The crash was caused by the ailerons being connected in reverse.

Bombing Philosophy

The 1930s was an interesting period in the development of the bomber aircraft. It was during this time that a great fear of huge fleets of bombers devastating entire cities, destroying public morale and almost winning wars on their own developed in many countries of the world, not just among the public and the politicians but some sections of the military as well. This 'bomber phobia' – and it should be mentioned that at this time no extensive strategic bombing of cities and other targets had as yet occurred – reached the point where international agreements restricting any nation's bomber force were called for, culminating in a series of discussions being held in Geneva in 1932 under the auspices of the League of Nations.

Opinions on the subject varied from British Prime Minister Stanley Baldwin advocating the abolishment of all bomber forces and the slowing down of the advancement of civil aviation through a reduction in funding, to more moderate views which suggested international control of civil aviation and the banning of bomber research and development.

Baldwin's extreme view was not atypical and reflected the general view in Britain at the time, although it should be noted that the British Air Ministry's opinion was contrary to the Prime Minister's. Baldwin's theory was that heavy bombing could cause the collapse of civil morale and lead to surrender even before a country's navy and army had been involved in the battle.

Bitter experience in the 1939-45 war proved the opposite to be the case. While strategic bombing of military and industrial targets could certainly be effective, the bombing of populated areas merely hardened the resolve of those civilians on the receiving end of it.

It was Adolf Hitler's coming to power in Germany in 1933 that began to change the previous opinion of the moral and military worth of bombing, at least as far as the British Govern-

Looking down on late production Lancaster I RE172, built in May 1945. Note the H2S radome under the rear fuselage.

ment was concerned. Here was a fascist regime which would obviously have no part of any international treaties dealing with bombers or any other military matter. Moreover, some within the British Government and military saw Hitler's Germany as a potentially direct threat to Britain, and advocated a strengthening of the armed forces in general.

The Royal Air Force was a fairly run down organisation in the early and mid 1930s with a strength of fewer than 600 aircraft, most of which were at best obsolescent. France's air force was three times as strong, and by 1935 Nazi Germany's *Luftwaffe* was already stronger than the RAF (and growing every day) with generally more modern aircraft in service.

In the face of this and the growing uncertainty over Germany's intentions, several RAF expansion schemes were put into place, the object being to triple the size of what had been the world's largest air force in 1918 and to introduce more modern aircraft. Even so, the RAF introduced a new biplane fighter type – the Gloster Gladiator – as late as 1937 and it would not be until the very end of that same year that the monoplane Hawker Hurricane began operational service with a squadron. The Supermarine Spitfire joined its first squadron in July 1938, just over a year before Britain and Germany were at war.

In 1935 the RAF's situation with bombers was even worse, a mere three squadrons of biplane Handley Page Heyfords comprised the 'heavy' bomber force backed by reasonable

quantities of small, single engined biplanes of the Hawker Hart/Audax/Demon family. The lumbering Heyford was scarcely an aircraft of the future with its open cockpit, fixed undercarriage, top speed of just 142mph (228km/h) when lightly loaded and considerably less when bombed up, a bomb load of up to 4,000lb (1,815kg) carried over extremely short distances or a modest 1,650lb (750kg) over 900 miles (1,480km) and a defensive armament of just three 0.303in Lewis machine guns. As excellent as Handley Page's series of bomber biplanes had been in World War I and during the 1920s, by 1935 a new direction was required.

That new direction – metal monoplanes with retractable undercarriage and more substantial defensive armament – was already being set in other countries. By the middle of 1935 Boeing had flown the prototype of its Model 299 (later B-17 Flying Fortress) four engined bomber; in Germany the Dornier Do 17 and Heinkel He 111 twin engined medium bombers were being tested and the first Junkers Ju 88 was just over a year away.

Between 1935 and the outbreak of war in September 1939 several new British bomber designs had reached squadron service, among them the medium Armstrong Whitworth Whitley and more successful Handley Page Hampden and the medium/heavy Vickers Wellington, the aircraft which bore the brunt of Bomber Command's operations until the four engined 'heavies' – notably the Handley Page Halifax and Avro Lancaster – arrived in strength.

The ancestor of the line, an early production Manchester I (L7284 of 207 Squadron RAF) with original short span tailplane, small endplate fins and added central dorsal fin. The prototype Manchester flew in July 1939. (via Neil Mackenzie)

(above) No 207 Squadron was the first operational unit to fly the Manchester, receiving its aircraft from late 1940. Illustrated is Manchester I L7380, complete with mid upper turret fitted. (via Neil Mackenzie)

(below) Another 207 Squadron Manchester, this time a definitive Mk.IA with longer span tailplane, larger endplate fins and the dorsal fin deleted. Ongoing problems with the Rolls-Royce Vulture engine restricted Manchester production to 201 examples including prototypes. By mid 1942 the aircraft had been taken off operations as the Lancaster began to arrive in numbers. (via Neil Mackenzie).

To Lancaster via Manchester

Two specifications for heavy bombers were issued by the Air Ministry in 1936. The first (Specification B.12/36) ultimately resulted in the Shorts Stirling, the RAF's first four engined heavy bomber. The Stirling entered service in 1940 but was quickly outclassed by the Lancaster and Halifax due to its poor altitude performance. This resulted from the original specification limiting its wing span to 100 feet (30.4m) so as to ensure compatibility with what was stated to be the maximum door width of RAF hangars of the time.

Specification P.13/36 indirectly led to the Lancaster (via the Manchester) and the Handley Page Halifax. The predecessors of both these heavy bombers were intended to be powered by a pair of 1,760hp (1,315kW) Rolls-Royce Vulture 24 cylinder (in X-form) engines, the result of the pairing of two Rolls-Royce Peregrine V12 cylinder blocks to a common crankshaft. The complicated Vulture was a troublesome and unreliable engine, Handley Page switching from two of them in its proposed HP.56 to four Rolls-Royce Merlin in 1937 – before its design had flown – thus dropping out of the P.13/36 competition but creating the Halifax. Avro built the Manchester with two Vultures first, put the aircraft into production and at the same time modified it to take four Merlins as the Lancaster.

Specification P.13/36 called for a fairly advanced aircraft, the issue of a production specification depending on the successful outcome of trials. The specification was really for a medium/heavy bomber and required a fast and long ranging aeroplane capable of carrying a 4,000lb (1,815kg) bomb load over a range of 3,000 miles (4,830km) at a speed of 275mph (442km/h). Powered gun turrets were

Two views of the first Lancaster prototype (BT308) as originally flown on 9 January 1941 with Manchester I short span tailplane and triple fins. This aircraft was powered by 1,075hp (800kW) for takeoff rated Merlin X engines, while production aircraft had the Merlin 20 and its derivatives.

to be installed in the nose and tail (and later in the mid upper position as well) while one novelty was the requirement for the aircraft to be launched with the aid of a catapult, not unlike that used on aircraft carriers. The theory was that higher takeoff weights could be achieved using a smaller wing. The idea was tested on the prototype Manchester but not proceeded with.

The Avro Type 679 Manchester was by far the largest aircraft built thus far by a company whose previous efforts had been small to medium sized aeroplanes. Roy Chadwick's design bristled with what was ad-

vanced technology in its day. An all metal mid winged monoplane with twin endplate tail fins, retractable undercarriage and provision for three powered turrets housing a total of eight 0.303 Browning machine guns, the Manchester would carry a crew of seven and a maximum bomb load of 10,350lb (4,700kg) in a massive bomb bay 33 feet (10 metres) long.

This bomb bay would prove to be very useful later on in the Lancaster when ever larger bombs needed to be carried including the 22,000lb (10,000kg) 'Grand Slam' which at a length of more than 25 feet (7.7 metres) could only be accommodated by a Lancaster and then with the bomb doors removed, portions of the bay cut away and the weapon left partially exposed to the slipstream.

The prototype Manchester (RAF serial number L7246) took to the air for the first time on 25 July 1939 from Manchester's Ringway Airport with Captain H A ('Sam') Brown at the controls. Several problems became immediately apparent. Apart from those associated with the Vulture engine's consistent tendency to overheat and to seize (then catch fire) as a result of lubrication problems, the airframe also needed some work. A lack of directional stability was temporarily fixed by fitting a stubby central vertical fin (two styles were tested) but it quickly became obvious that a larger wing was going to be needed.

The second Lancaster prototype (DG595), first flown on 13 May 1941. This aircraft was more representative of production B.Is with Merlin 20s, the definitive tailplane/fin arrangement and armament.

The fourth production Lancaster B.II (DS604) in 61 Squadron markings. This version was powered by Bristol Hercules radial engines and was developed in case of a shortage of Merlins. There wasn't, so production was limited to 300 aircraft, all by Armstrong Whitworth Aircraft. (via Neil Mackenzie)

The Manchester's original 80ft 2in (24.43m) wing span and the area it produced would have been adequate had the proposed catapult launching system been used, but without that more lift was required. This was provided by the fitting of modified tips to the second prototype (L7247) which increased the wing span to 90ft 1in (27.45m). The second prototype first flew in May 1940 and was much more representative of a production Manchester. Apart from the extended wings it featured the third fin and more complete operational equipment including turrets and guns.

Despite the problems with its powerplant, production Manchester Mk.Is began to reach the first RAF squadron to equip with the type – No 207 – late in 1940. By this time orders for 600 Manchesters had been placed with Avro, Metropolitan-Vickers, Fairey and Armstrong Whitworth but in the event only 201 were built in-

cluding the prototypes. Of those, 158 came out of Avro and the remainder from Metrovick with the other two companies failing to produce any before the decision to replace the aircraft on the production lines with the Lancaster was made.

Early Manchesters were built to Mk.I standards and later aircraft featuring redesigned and greater span horizontal tail surfaces coupled with larger endplate fins were designated Mk.IAs. This removed the directional stability problem once and for all. Additionally, other problem areas – like the complex and in many ways innovative hydraulic system and a weak undercarriage – were now fully proven, but the Vulture's woes continued.

Although rated at 1,760 horsepower (1,315kW), the engine failed to deliver that, and in order to induce some kind of reliability it was derated further to produce just 1,500hp (1,120kW) in service. This in turn re-

sulted in lower maximum takeoff weights being permitted, the specified 56,000lb (25,400kg) turning out to be 50,000lb (22,680kg) in practice. This had an obvious effect on the Manchester's payload/range characteristics.

By the middle of 1942 the Manchester had been taken off operations and relegated to Operational Conversion Units; within a year even those units were getting rid of them as engine related crashes began to take their toll.

Two proposals to rescue what was obviously a sound airframe design were made in 1939, a year before the first Manchesters were delivered to the RAF. The first was the proposed Manchester II with either Napier Sabre inline or Bristol Centaurus radials; the second involved putting four Rolls-Royce Merlins onto a further extended wing and calling the aircraft the Manchester Mk.III.

This rear view of a Lancaster shows some interesting detail such as the dihedral of the out wing sections in contrast to the centre section. The tailplane's span was just over 33 feet or 10m. (via Neil Mackenzie)

More With Four

The Sabre and Centaurus proposals were shortlived, efforts quickly concentrating on the Manchester Mk.III with four Merlins. Design work was carried out under the Avro Type Number 683, the design retaining the Manchester's fuselage, tail unit and centre wing, the latter combined with new outer panels of increased span and the structural changes necessary to accommodate four rather than two engines. The original Vulture engine mount positions were modified to accept the inboard Merlins while new mounts were incorporated for the outboard engines. The Merlins would be installed as a complete 'power egg'.

Avro's early estimates had the Type 683 capable of carrying a 12,000lb (5,443kg) bomb load over a distance of 1,350 miles (2,172km) or 8,000lb (3,628kg) over 2,000 miles (3,218km) at the most economical cruising speed of 190mph (305km/h). Using a higher cruising speed of 245mph (394km/h) reduced the range figures by about 25 per cent.

This was considered to be useful performance and in view of that and the problems being suffered by the Manchester, the Air Ministry decided

in November 1940 to stop production of the Manchester after the 200th aircraft and replace it with the Avro 683 'Manchester III' at the various factories and production programme which had been built around the earlier aircraft. Interestingly, the name 'Lancaster' had been associated with the project within Avro for a time, but the company was told to stop referring to it as such for security reasons.

In the meantime, an order for four prototype Lancasters had been placed (that for the first in September 1940) of which only three (BT308, DG595 and DT810) were built, the fourth (DT812) being cancelled. BT308 and DG595 were powered by Merlins while DT810 had Bristol Centaurus radial engines.

The differences between the Manchester and early Lancasters were kept to a minimum, the first 400 of the 7,377 ultimately built up to early 1946 starting out as Manchesters in the factory and emerging as Lancasters. The obvious differences were the installation of four Rolls-Royce Merlin liquid cooled engines equipped with two-speed single stage superchargers, an increase in wing span to 102ft 0in (31.09m), ex-

tra fuel capacity (2,154 imp gal/9,792 litres) in all but the very earliest examples with provision for auxiliary tanks in the bomb bay and vastly increased carrying capacity and payload/range capability. The maximum takeoff weight of a Lancaster was normally 68,000lb (30,844kg) with 72,000lb (32,660kg) permissible later in the war as a maximum overload.

There were obviously differences associated with the installation of four, rather than two engines, but basically the Lancaster was indeed a 'Manchester Mk.III'. This high degree of commonality, combined with an uncomplicated airframe design which lent itself to assembly by the basically unskilled hands who built the aircraft in very large numbers, allowed a rapid changeover from Manchester to Lancaster production after the decision to do so had been made.

This factor also allowed the first prototype (BT308) to be built and flown very quickly, only four months lapsing between go ahead and first flight on 9 January 1941. Like its predecessor, the first Lancaster recorded its maiden flight at Manchester's Ringway Airport in the hands of Captain H A Brown.

Lancaster front end detail showing undercarriage, nose/cockpit section and engine cowling. (via Neil Mackenzie)

This aircraft differed from production Lancasters in that it was fitted with a Manchester Mk.I tail unit with the smaller span horizontal surfaces, small vertical surfaces and the third vertical fin. It was powered by four Merlin X engines (the first version with two speed/single stage supercharging) producing 1,075hp (800kW) for takeoff and 1,265hp (945kW) at 17,000 feet.

Initial production Lancasters would be fitted with Merlin XX (subsequently given the Arabic appellation '20') engines offering 1,280hp (955kW) at takeoff while more powerful variations of the Merlin 20 were used throughout the production run with 'tailoring' of the two speed supercharger's gear ratios providing different power ratings at different heights. This was a characteristic of the two speed/single stage and later two speed/two stage Merlins; the heights at which maximum power was achieved could be adjusted by changing the supercharger gearing. As a result, some Merlins were at their best at low altitudes, some at high altitudes and others in the middle.

There was little doubt that the installation of four Merlins would transform the aircraft, but there was still some doubt about the availability of Merlins as the demand for it was exceptionally high. This is why an alternative powerplant (the Bristol Hercules radial) was specified for one of the prototypes but the situation was greatly relieved in September 1940 with the signing of an agreement which would see Merlins being manufactured in large numbers by Packard in the USA. Packard moved quickly. Its first Merlin was running just under a year later and full production began in 1942.

Into Production

After completing initial manufacturer's trials, the first prototype was presented to the Aircraft & Armament Experimental Establishment (A&AEE) at Boscombe Down in late January 1941 for preliminary trails. It received favourable reports as to its handling and especially its speed, a maximum of 310mph (499km/h) at 21,000 feet being recorded. As initially flown, BT308 retained the Manchester's short span tailplane, centre dorsal fin and nose (Frazer Nash FN 5) and tail (FN 4A) gun turrets. By March the definitive longer span with enlarged endplate fins tail unit had been fitted.

This historic aircraft then underwent further trials at Boscombe Down before being used to visit some of the operational squadrons which would soon be operating the new bomber. It was delivered to Rolls-Royce for exhaust flame damp-

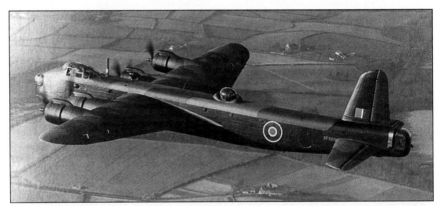

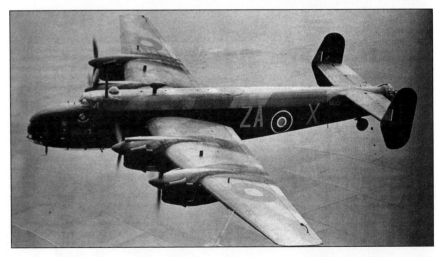

The RAF's three heavy bombers of World War II (top to bottom): Avro Lancaster, Short Stirling and Handley Page Halifax. Of the trio, the Lancaster was notably the most effective. The Lancaster illustrated is B.I R5857/OL-F of 83 Squadron in 1942.

ing trials in early 1942 then to the Royal Aircraft Establishment (RAE) at Farnborough for trials with the Metropolitan-Vickers F.2 turbojet which was installed in the tail. It was struck off charge in May 1944.

The second prototype (DG595) first flew on 13 May 1941 and was representative of the production Lancaster B Mk.I with Merlin 20s, the definitive tailplane/fin arrangement, larger mainwheels and production standard gun armament comprising a two gun FN 5 nose turret, four gun FN 20 tail turret, two gun FN 50 dorsal (mid upper) turret and two gun FN 64 ventral turret, the latter appearing on early production aircraft but disappearing after that. All the turrets were equipped with 0.303in Browning machine guns.

The third prototype (DT810) was flown on 26 November 1941, nearly four weeks after the first production

Lancaster had taken to the air. In effect the prototype for the Lancaster Mk.II, DT810 was powered by four Centaurus radial engines, this version being developed as insurance against a possible shortage of Merlins. Following trails with the A&AEE, RAE and Bristol, this aircraft was struck off charge in 1944.

The first production Lancaster (L7527) was flown on 31 October 1941 and the RAF's No 44 (Rhodesian) Squadron became the first operational unit to receive the new bomber, its aircraft beginning to arrive in December in preparation for the first operational sortie on 3 March 1942. This was a minelaying task and the Lancaster's first bombing mission was recorded exactly one week later when two 44 Squadron aircraft joined a raid on Essen.

Although the Lancaster's development period had been quick and successful and its subsequent career a triumph, the aircraft was not without some flaws. Early problems with the fuel system's immersed pumps kept a large part of the fleet on the ground in 1942 while failures of the wingtips under high load briefly grounded the entire fleet in the same year. Fin failures under high loads (such as when

violently manoeuvring to evade enemy fighters) required some strengthening of the tailplane/fin junction and some high speed dives resulted in the fabric on the elevators separating and causing an uncontrollable dive into the ground.

Five companies were involved in the production of the three main Lancaster variants in Great Britain: Avro, Metropolitan-Vickers, Armstrong Whitworth, Austin Motors and Vickers-Armstrong. Between them they built 6,947 Lancasters to which must be added 430 built in Canada by Victory Aircraft, a company formed by the National Steel Company of Canada especially for that purpose.

A feature of the Lancaster variants was – with one exception – their basic similarity, although the level of detail differences (in equipment fits and running modifications installed) could be quite substantial. The Lancaster B.I was the basic model with Rolls-Royce built Merlins. A total of 3,437 was manufactured including three prototypes.

The Lancaster B.II represented the only major departure from the basic formula in that 1,650hp (1,230kW) Bristol Hercules VI or XVI 14-cylinder radial engines were fitted. The Her-

cules-Lancasters had a lower service ceiling and higher fuel consumption than the original but offered slightly greater speed. The availability of the Packard Merlin from 1942 removed the need for an alternative and only 300 Lancasters IIs were built, all of them by Armstrong Whitworth in 1942-43.

The Lancaster III was simply a Mk.I fitted with Packard Merlins which were entirely compatible with airframes originally designed to take Rolls-Royce Merlins. The differences were very minor ones associated with some items of American equipment which came with those engines. Lancaster III production amounted to 3,030 aircraft. The Canadian built Lancasters were similar to the Mk.III but were designated Mk.X to differentiate.

A fourth Lancaster variant appeared late in the war, the Mk.VII. This was equivalent to a late model Mk.I but with tropical equipment for service in the Far East and the replacement of the mid-upper and rear 0.303in machine guns with 0.50in (two in each turret) guns. Austin Motors built 180 Mk.VIIs in 1945. Other Lancaster Marks which appeared were conversions of the basic models and all are discussed in more detail in the following chapter.

The power of the World War II heavy bomber illustrated by an example of the war's most destructive conventional bomb – the 22,000lb 'Grand Slam' – about to be loaded aboard a Lancaster I (Special) of the famous 617 (Dambusters) Squadron. This photograph was taken in March 1945. (via Neil Mackenzie)

Aeronavale (French Navy) Lancaster B.VII WU-15 photographed at Bankstown Airport, Sydney Australia in late 1964/early 1965 prior to its flight to England for preservation. After several changes of ownership over the years, the Lanc is now in the safe hands of the Lincolnshire Aviation Heritage Centre, East Kirkby, back in its original RAF markings as NX611. (Eric Allen)

Lancaster WU-15/NX611 at Bankstown. For its epic flight to Britain in 1965 it carried the registration G-ASXX. The Lanc's final flight was recorded in June 1970. (Eric Allen)

An unusual overhead view of the Battle of Britain Memorial Flight's Lancaster B.I PA474. This famous Lanc has been flying with the BBMF since 1968 and attracts enormous amounts of attention wherever she goes.

Another preserved Lancaster is B.I W4783/AJ-G (top) – a 90 mission veteran – at the Australian War Memorial, Canberra. Not quite so well preserved when photographed at Oshawa, Ontario in 1971 is this RCAF Lancaster 10 KB889 (bottom). Happily, this aircraft is now being well looked after by the Imperial War Museum at Duxford in the UK. (Eric Allen)

LANCASTER VARIANTS AND DEVELOPMENT

LANCASTER B.I

The Lancaster I was the first production variant of the Avro bomber and formed the basis for all which followed. It was the most produced version (3,434 built) and as noted, was for all intents and purposes the same aircraft as the Lancaster III (3,030 built), the latter differing only in having Packard Merlins installed instead of Rolls-Royce built engines.

Due to this similarity, the development of the Lancaster's armament and operational equipment during the course of World War II applies equally to both versions.

The first production Lancaster B.I (L7527) was flown on 31 October 1941 from Avro's Woodford (Manchester) facility and by the end of that year a further 12 aircraft had been completed. Production was spread between seven sites, Lancaster Is emerging from Avro Woodford (840), Avro Yeadon (54) Metropolitan-Vickers Woodford (921), Armstrong Whitworth Coventry (911), Austin Motors Birmingham (150), Vickers-Armstrong Castle Bromwich (300) and Vickers-Armstrong Chester (258). The first four sites were all producing Lancasters in large numbers by the end of

1942. A further 200 Lancasters were ordered from Short Bros and Harland in Belfast, but these were cancelled.

Production built up rapidly during 1942, from 23 per month in January to 91 per month by December. The peak production rate was achieved in the third quarter of 1944 when 260 Lancasters of all models were pouring from the various factories each month.

Gradual Development

Very early production Lancaster Is differed from the vast majority in several details. They were fitted with the ventral gun turret which was soon removed in service and production (leaving a blind spot which would subsequently be exploited by *Luftwaffe* night fighters); they had a fuel capacity of 1,710imp gal (7,774 l) in four wing tanks; power was provided by four Merlin 20s rated at 1,280hp (955kW) for takeoff and 1,260hp (940kW) at 11,750ft; the mid upper turret was unfaired and the guns unrestricted in their depression; and maximum weight was set at 61,500lb (27,896kg).

What may be termed the 'standard' Lancaster I quickly followed with

higher boosted Merlin 22 engines giving 1,390hp (1,035kW) for takeoff and 1,460hp (1,090kW) at 6,250ft; six fuel tanks in the wings giving a total capacity of 2,154imp gal (9,792 l); a normal maximum takeoff weight of 65,000 to 68,000lb (29,484 to 30,845kg); and a faired mid upper turret combined with a gun taboo track, the combination reducing drag slightly and preventing gunners from shooting their own Lancasters' tails off!

The usual crew complement on a Lancaster was seven: pilot, flight engineer, radio operator, navigator, bomb aimer/gunner and two other gunners. The few aircraft equipped with the ventral gun turret would carry an extra crewmember.

Further powerplant development allowed the Lancaster to operate at higher weights and carry the very big bombs only it was capable of lifting. The Merlin 24 engine featured a further increase in maximum boost and was capable of producing 1,620hp (1,210kW) for takeoff, 1,640hp (1,225kW) at 2,000ft and 1,500hp (1,120kW) at 9,500ft. This allowed a maximum weight of 68,000lb (30,844kg) or in exceptional circumstances – such as when carrying the 22,000lb (9,979kg)

One of the best known Lancasters, B.I R5868/PO-S 'S for Sugar' of 467 Squadron RAAF. This aircraft is now preserved at the RAF Museum, Hendon. It is shown here being bombed up with a 4,000lb (1,815kg) 'Cookie' and smaller bombs. (via Neil Mackenzie)

Looking down on a Lancaster, highlighting its long and efficient wings, perhaps the secret of the aircraft's substantial lifting power. (via Neil Mackenzie)

'Grand Slam' deep penetration bomb – 72,000lb (32,660kg).

The Lancaster's standard propeller was a de Havilland Hydromatic constant-speed and fully feathering three bladed unit, while a Nash Kelvinator unit with paddle blades could also be fitted. This improved performance at higher altitudes at the cost of some performance lower down.

The Lancaster I's nose guns installation remained constant throughout its production life, but the rear turret in particular evolved due to changing operational circumstances. The original Frazer Nash FN 20 turret with four 0.303in Browning machine guns was eventually replaced by an FN 82 or in some cases Rose-Rice unit, both with two American 0.50in Browning heavy machine guns which offered more effective defensive fire. Lancaster rear gunners also found themselves removing some perspex from their turrets so as to improve the view and minimise the effects of ice and mist.

Some Lancaster Is which were intended to be built as Mk.VIIs (see below) with Martin dorsal turrets mounted further forward were completed with a standard FN 50 in the new position due to a shortage of the American unit. Others had Frazer Nash FN 79 dorsal turrets fitted with a pair of 0.50in guns.

The Payload

Probably the Lancaster's major asset was its enormous bomb bay. Nearly 33 feet (10m) in length, the bay allowed great flexibility in bomb load and permitted the aircraft to carry bigger weapons internally (and semi externally with modification) than any other bomber of the era.

With standard bomb bay doors the Lancaster could carry the 4,000lb (1,814kg) 'Cookie' bomb internally, while bulged doors gave the capability to carry an 8,000lb (3,629kg) bomb (plus other smaller bombs) or a single 12,000lb (5,443kg) 'Tallboy' spin stabilised deep penetration bomb for use against reinforced targets such as submarine pens. The 22,000lb (9,979kg) 'Grand Slam' installation is discussed below.

The normal maximum bomb load for a Lancaster was 14,000lb (6,350kg), comprising similar or mixed loads, depending on the mission. Some representative loads might comprise: 14 1,000lb (454kg) bombs for industrial demolition; one 4,000lb (1,814kg) and three 1,000lb (454kg) bombs plus six incendiary containers for blast, demolition and fire; one 4,000lb (1,814kg) and 18 500lb (227kg)

Early production Lancaster I R5556/KM-C of 44 Squadron, the first RAF operational unit to fly the type. Note the ventral gun, soon deleted from all but a few aircraft. (via Philip J Birtles)

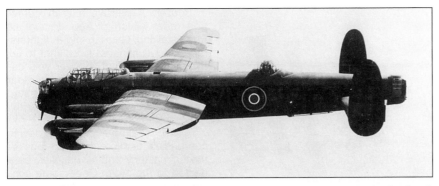

Lancaster I R5689/VN-N of 50 Squadron RAF. This aircraft is from an early production batch delivered in mid 1942. (via Neil Mackenzie)

bombs for carpet bombing of tactical targets; six 2,000lb (907kg) armour piercing plus three 500lb (227kg) bombs for use against docks, ships and fortifications; or up to six 1,850lb (839kg) parachute mines for minelaying operations. Many other combinations of bomb load were possible.

Black Boxes

A significant part of the Lancaster story is the pioneering role it played in the deployment of electronic offensive, defensive, navigation and countermeasures equipment. Great technological advances were made by both sides over the last three years of the war, with Germany's night fighter force incorporating ever improving radar equipment designed to track the bombers, which in turn were equipped with more and more devices intended to help them find their targets and to thwart the defending fighters. A brief description of the major items used operationally in Lancasters follows:

Gee: Available for fitting to Lancasters when squadron service began in early 1942, Gee was an effective radio navigational device which was accurate enough to find a target although it could not be used for precision blind bombing. Gee worked by the transmission of two circular radio pulses and in combination with a receiver incorporating an indicator unit and special charts, a navigator could plot his aircraft's position with ac-

ceptable accuracy. The equipment was not easy to jam and was protected by various deception operations.

H2S Radar: The world's first ground mapping radar, based on the existing ASV (air-surface vessel) radar, H2S 'painted' a rough radar picture of the terrain below on a cathode ray tube screen at the navigator's station. Flat terrain or sea returned no

signals but coastlines, towns and the like created an image on the screen. With the aircraft's position in the centre of the screen, the navigator could read the display much like a map, regardless of cloud cover, and use it for blind bombing.

The rotating antenna was mounted in a radome under the rear fuselage (where the ventral gun position had once been) and improved versions were fitted to just about all RAF bombers as the war progressed.

H2S became available in early 1943 and was first allocated to the Pathfinder Force's Lancasters with others following as the equipment became more readily available. Although H2S had numerous operational benefits, it also had one large disadvantage in that being an active radar, its signal could be picked up by German *Naxos Z* equipment, thereby helping to lead the fighters to the bomber. As a result, the H2S normally had to be turned off except when needed.

Monica: A rearwards looking radar developed from that fitted to early British night fighters. Giving an audible warning when it detected an aircraft within a 45 degrees arc behind the bomber, its detection range was up to about 900 metres and the 'bleeping' signal it produced increased in frequency as the encroaching aircraft closed in.

Monica's disadvantages were a tendency to give off false alarms

Another early Lancaster I, this time W4113 in the markings of No 1661 Conversion Unit.

when other bombers crossed behind – not uncommon with large numbers occupying one piece of sky – and the fact that like H2S it was an active radar and could be used by German night fighters to home in the bomber. This is exactly what happened via the German *Flensberg* equipment, with the result that Monica was of little use after a short period. Germany also developed equipment which could home in the bombers' identification friend or foe (IFF) equipment.

Lancaster Is PB434 and PB436 with H2S radar installed. These Lancasters are from a Woodford built batch delivered July 1944. (via Neil Mackenzie)

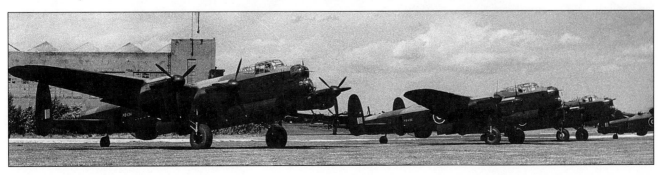

Boozer: A passive aircraft detection radar which picked up signals from stalking German night fighters' radar and illuminated a red light to tell the pilot of the fact. Boozer was capable of discriminating between radar transmitters: a yellow light illuminated if the Lancaster was being tracked by radar on the ground.

Tinsel: A simple jamming device comprising a microphone in one of the Lancaster's engine nacelles connected to the radio. Once some conversation between a German controller and a night fighter had been monitored, the radio would be tuned to the appropriate frequency whereupon the conversation would be drowned out by the sound of a roaring Merlin!

Window: A simple but very effective radar jamming device consisting of strips of aluminium foil cut to a specific length and dropped from bombers in large quantities. Falling to the ground, the Window would present many false returns to the German radar, in effect blotting it out and hiding the true position of the bombers.

Window proved to be generally effective and although it was ready for use by the end of 1942, its deployment was delayed unit July 1943 as it was thought exposure to the enemy would result in it being used by them.

'Tail End Charlie' in a Lancaster. The aircraft's rear gun evolved throughout its operational life, starting with four 0.303in machine guns in a Frazer Nash FN 20 turret and progressing to an FN 82 or in some cases Rose-Rice unit, both with a pair of 0.50in guns. The removal of some of the perspex for better visibility and to minimise the effects on ice and mist became a standard modification.

As more power became available from the Rolls-Royce Merlin engine, Lancasters were fitted with paddle bladed Nash Kelvinator propellers to transmit it more effectively. This Lancaster is about to receive its night's payload. (via Neil Mackenzie)

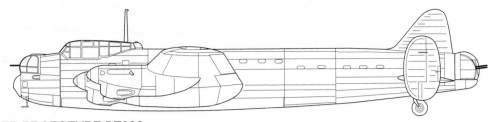

LANCASTER PROTOTYPE BT308

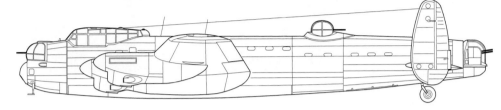

LANCASTER B. Mk I (Early)

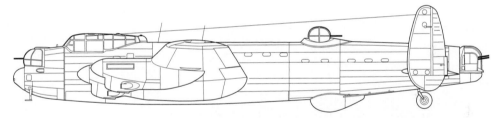

LANCASTER B. Mk II

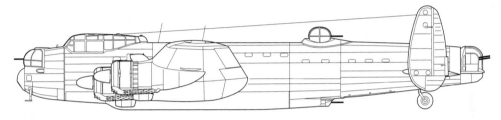

LANCASTER B. Mk III (Late)

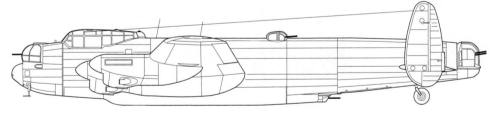

LANCASTER B. Mk X

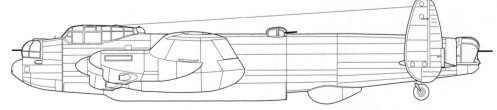

LANCASTER B. Mk VII (FE)

Juanita Franzi 1995

The pressure of mounting bomber losses resulted in the decision to finally use Window, although there is an irony in the fact that Germany had also developed a similar system but had decided not to use it for the same reason!

Most raids from the second half of 1943 had Window involved, the material being dropped from a container in the Lancaster's lower starboard nose by the bomb aimer.

Mandrel: For jamming German *Freya* early warning radar by transmitting interference on the appropriate frequencies.

Airborne Cigar (ABC): Another jamming device and perhaps indicative of the increasing 'trickery' being used by both sides as the technology became available. ABC was basically a radio receiver tuned to German night fighter voice control frequencies. An additional German speaking crewmember aboard the Lancaster would monitor these frequencies and then on the discovery of communication switch on a powerful transmitter

Navigating by black box. A Lancaster navigator consults his Gee indicator screen.

on the detected frequency and wipe out the signal with interference.

Gee-H: A precise blind bombing aid which entered service in late 1944, Gee-H permitted accurate attacks even through heavy cloud cover by using a transmitter-receiver

in the aircraft to accurately measure distance from ground beacons. It was introduced into service at the same time as the much improved H2S Mk.VI radar and that in combination with *Oboe*, a highly accurate radio marking aid carried by Pathfinder Force Mosquitos and the Lancaster's Mk.XIV gyroscopically stabilised and computer linked bombsight (available from late 1942) made for a degree of bombing accuracy undreamt of earlier in the war.

Lancaster B.I (Special)

Devised especially to carry the largest bomb of World War II – Barnes Wallis' 22,000lb (9,980kg) 'Grand Slam' deep penetration weapon – the Lancaster B.I (Special) differed from the standard aircraft by incorporating several modifications which enabled the weapon to be carried. No other bomber of the era was capable of lifting and dropping the Grand Slam.

In order to accommodate the bomb, the Lancasters' bomb doors were removed and the bay faired in at both ends. The weapon hung partially exposed to the airstream and in order to save weight most Specials had their nose and mid upper turrets removed and the tail turret armament reduced to just one pair of guns. Other equipment was also removed to save weight, resulting in an empty weight of about 36,000lb (16,345kg). An increased maximum weight of 72,000lb (32,660kg) was permitted, allowing the Lancaster Special to lift its own weight in fuel and payload. More powerful Merlin 24 engines were installed and the undercarriage was strengthened.

Thirty-two Specials were built at Avro's Woodford factory in the first three months of 1945, the variant flying only with the RAF's 617 'Dambusters' Squadron.

Lancasters in production. Seven factories in Britain and one in Canada were involved in the Lancaster manufacturing programme, between them delivering 7,377 aircraft.

Lancaster Is of 35 Squadron with the bulged bomb bay doors which allowed the carriage of an 8,000lb (3,630kg) bomb internally. These postwar aircraft are from the last Lancaster production batch, delivered from Armstrong Whitworth at Coventry between October 1945 and March 1946.

Barnes Wallis developed the idea of the deep penetration bomb during 1943, the basic principal being that the spin stabilised weapon would reach a supersonic terminal velocity (when dropped from sufficient height) and bury itself deep into the ground before exploding. From this they earned the nickname 'earthquake bombs'. The main targets for such a weapon would be underground and reinforced structures such as U-boat or missile pens.

The first expression of the 'earthquake' bomb theory was the 12,000lb (5,443kg) 'Tallboy', first used by 617 Squadron in June 1944 against a railway tunnel in France and subsequently against other targets including the

Lancaster B.I Special PB995 of 617 Squadron in flight with Grand Slam bomb attached. (via Neil Mackenzie)

Thirty-two Lancaster B.I Specials were produced in 1945, specifically to carry the 22,000lb (10,000kg) Grand Slam 'earthquake' bomb. The bomb can be seen cradled under the aircraft, partially exposed to the airflow. In order to accommodate the weapon, the Lancasters' bomb doors were removed and the bay faired in at both ends. The nose and mid upper turrets were removed to save weight. (via Neil Mackenzie)

One of the pair of Lancaster Is experimentally fitted with a 1,200imp gal (5,455 l) saddle tank in 1944. Aircraft so equipped were intended to operate in the Far East against Japan but were not needed. The experiment was unsuccessful anyway, the modification producing handling difficulties.

celebrated raid which resulted in the sinking of the German battleship *Tirpitz* in Norway in November 1944.

Only 41 'Grand Slam' bombs were dropped in the last three months of the European war, mostly against U-boat pens but also in March 1945 against the Bielefeld viaduct, with considerable effect.

Lancaster B.I (FE)

Twenty-five new production Lancaster B.I (FE)s were built by Armstrong Whitworth at Coventry in mid 1945, intended for use with Tiger Force in the Far East in the war against Japan. Before this became unnecessary with Japan's early capitulation (due to the dropping of atomic bombs on Hiroshima and Nagasaki), plans were in place to send several hundred Lancasters in FE configuration and the new Lincoln to the Pacific area.

Because of the great ranges involved, flight refuelling of the aircraft was investigated, as was the installation of a bulbous 1,200imp gal (5,455 l) saddle tank on top of the fuselage, their fairings encasing the top of the cockpit canopy. Two Lancaster Is (HK541 and SW244) were converted to test the idea in 1944, but poor aerodynamic characteristics resulted in handling problems and the idea went no further. See also the Lancaster B.VII (FE) below.

Lancaster PR.1

A postwar photographic reconnaissance conversion which served only with the RAF's Nos 82 and 683 Squadrons, the Lancaster PR.1 had all its turrets removed and faired over and cameras were installed in the bomb bay. The PR.1 was used for a comprehensive aerial survey of east central and west Africa between 1946 and 1952.

A PR.1 (PA427) was the last Lancaster operated by RAF Bomber Command, retired in December 1953; the very last Royal Air Force Lancaster in service was an MR.3 which survived until October 1956.

LANCASTER B.I

Powerplants: Four Rolls-Royce Merlin 20, 22 or 24 liquid cooled, V12 piston engines with two speed/one stage superchargers rated at (20) 1,280hp (955kW) for t-o and 1,160hp (865kW) at 20,750ft, (22) 1,390hp (1,035kW) for t-o and 1,435hp (1,070kW) at 11,000ft, (24) 1,620hp (1,210kW) for t-o and 1,500hp (1,120kW) at 9,500ft. De Havilland 5140 or Nash Kelvinator A5/138 three bladed constant-speed and feathering propellers of 12ft 0in (3.66m) diameter. Fuel capacity 2,154imp gal (9,792 l) in six wing tanks, provision for one or two 400imp gal (1,818 l) auxiliary tanks in bomb bay.

Dimensions: Wing span 102ft 0in (31.09m); length 68ft 10in (20.98m); height 20ft 4in (6.18m); wing area 1,297sq ft (120.49m²); wheel track 23ft 9in (7.24m); tailplane span 16ft 8.5in (5.09m).

Weights: Empty equipped 41,000lb (18,598kg); normal maximum 68,000lb (30,845kg); max overload (B.I Special) 72,000lb (32,659kg).

Armament: (Normal defensive) Frazer Nash FN 5 nose turret with two 0.303in machine guns and 1,000rpg; Frazer Nash FN 50 dorsal turret with two 0.303in machine guns and 1,000rpg; Frazer Nash FN 20 rear turret with four 0.303in machine guns and 2,500rpg; (early aircraft) Frazer Nash FN 64 ventral turret with two 0.303in machine guns and 750rpg.

(Offensive) normal maximum bomb load 14,000lb (6,350kg) or single 22,000lb (9,979kg) Grand Slam deep penetration bomb on B.I Special.

Performance: (Merlin 24) Max speed 242kt (447km/h) at 4,000ft, 249kt (462km/h) at 11,500ft; cruising speed 188kt (347km/h) at 20,000ft; initial climb (max weight) 270ft (82m)/min; time to 20,000ft 41.6min; service ceiling (max weight) 20,000ft (6,096m); service ceiling (mean weight) 24,500ft (7,467m); takeoff distance to 50ft (max weight) 4,650ft (1,417m); range with 10,000lb bomb load and standard fuel 905nm (1,673km); range with 7,000lb bomb load and one auxiliary tank 2,330nm (4,310km).

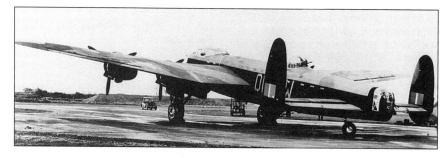

A pair of Lancaster IIs: DS604/QR-W (top) of 61 Squadron and DS771 (bottom) with bulged bomb bay doors. (via Neil Mackenzie)

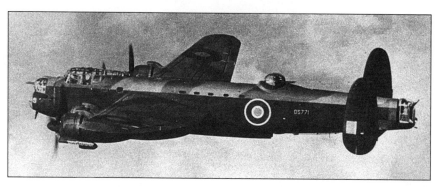

A Hercules engined Lancaster B.II cruises above the clouds, "after attacking flying bomb bases in Northern France", according to the official caption. Developed in case a shortage of Merlins developed, the B.II was not really needed and production was limited to 300 examples built by Armstrong Whitworth. (via Neil Mackenzie)

LANCASTER B.II

Concerns about a possible shortage of the very much in demand Rolls-Royce Merlin engine resulted in development of the Lancaster B Mk.II powered by four Bristol Hercules VI or XVI 14 cylinder two row radial engines with two speed superchargers rated at 1,615hp (1,205kW) for take-off and 1,675hp (1,250kW) at 5,000 feet. The first Hercules powered Lancaster to fly was the third prototype (DT210) which took to the air on 26 November 1941.

Production of the Lancaster II was undertaken by Armstrong Whitworth at Coventry, a total of 300 aircraft leaving that factory in 1942-43. Armstrong Whitworth switched to Lancaster I production after that, the changeover occurring in late 1943.

First deliveries were made to No 61 Squadron in September 1942 and only five other squadrons operated this variant: Nos 408, 426 and 432 (all Canadian) and No 514. The Lancaster II's first operational mission was performed in January 1943 when three 61 Squadron aircraft raided Germany.

Many Lancaster IIs featured the deepened bomb bay which enabled them to carry 8,000lb (3,630kg) bombs while many also had the ventral gun turret installed (or reinstalled), possible because none had H2S radar fitted. There were some detail changes to the Mk.II's specification during the course of its limited production run including replacing the original Hercules XVI engines with similarly rated XVIs.

Compared with the Merlin powered Lancasters, the B.II offered similar speeds and a better rate of climb at low-medium altitudes but fuel consumption was greater and altitude performance was poor by comparison, a fully loaded aircraft struggling to reach a ceiling of 15,000 feet (4,570m), some 4,000 feet (1,220m) less than a Lancaster I or III. Some 60 per cent of the 300 Lancaster IIs were lost on operations, many due to their inability to climb to a reasonable height.

As a result of this and the fact that the flow of Merlins from both British and American sources was adequate, the Lancaster II was removed from production at the end of 1943 and relegated to training duties at Heavy Conversion Units and other second line roles. It was declared obsolete in May 1945. The British squadrons which operated the Lancaster II switched to Merlin powered Mks.I and III, while the three Canadian units converted to Handley Page Halifaxes, all during 1944.

LANCASTER B.III

The arrival from 1942 of Merlins built in the USA by Packard ensured adequate supply of this vitally important powerplant and resulted in the creation of a new Lancaster variant, the B Mk.III. For all intents and purpose the Mk.III was the same aircraft as the Mk.I except for the use of Packard rather than Rolls-Royce Merlins. The differences were minimal (the use of US magnetos and carburettors among them) but regarded as sufficient to warrant a new mark number to differentiate for spares provisioning and administrative purposes.

The two Lancaster variants were manufactured side by side and their powerplants were theoretically interchangeable and although never built with a mixture of British and American engines installed, this was not uncommon in the field after engines had been changed following overhaul or damage. Lancaster Is often found themselves emerging from overhaul as Mk.IIIs due to engine changes, or *vice versa*!

In view of this similarity, the armament and equipment developments discussed in the Lancaster I section, apply equally to the Lancaster III.

Lancaster III production amounted to 3,030 aircraft, produced by Avro Woodford (2,135), Avro Yeadon (641), Metropolitan-Vickers (136) and Armstrong Whitworth (118). The first installation of a Packard Merlin in a Lancaster was made in August 1942 in Lancaster I R5849, the aircraft used by Rolls-Royce for tests, W4114 joining the programme the

H2S equipped Lancaster III PB410/OF-J of 97 Squadron, photographed in 1945. This Packard Merlin powered variant was indistinguishable from Mk.Is with their Rolls-Royce Merlins. (via Neil Mackenzie)

following month and regarded as the prototype B.III. Series production began in late 1942. The last Lancaster III (TX273) was delivered from Avro's Yeadon factory in October 1945.

The Packard Merlin

The chain of events which resulted in the Packard Motor Car Company's Aircraft Engine Division manufacturing the Merlin began in mid 1940 when Rolls-Royce was instructed to present the US Government with a complete set of Merlin drawings in case – as was possible at the time – Britain was overrun by Germany. In June 1940, Packard was asked if it would consider building 9,000 Merlin

XXs with two speed/one stage superchargers under licence, 6,000 for Britain (to power Hurricanes and Lancasters) and 3,000 for American use. A similar offer had already been made to Ford, which had turned it down due to perceived difficulties in modifying the engine to suit the company's production methods. Also, Henry Ford didn't think Britain would survive a German onslaught.

Packard accepted the challenge and by August 1940 Rolls-Royce engineers were in the USA helping the Americans prepare for production. An early decision was to build Packard Merlins with the newly developed two piece cylinder block and as a result of

it not being practical to immediately switch British production to this new feature, Packard Merlins were the first with it.

Packard did an extraordinary job, running its first two Merlin XXs (US designation V-1650-1) in August 1941, one year after development began. The Rolls-Royce engineers were amazed at how quickly Packard was able to work to the British company's tight production tolerances.

By 1942 the Packard Merlin was in full production, the American company eventually producing 55,523 engines between then and late 1945. This represented about 36 per cent of total Merlin production of some 155,000 units. Packard was producing 600 Merlins per month by July 1942, increasing to 1,300 per month a year later and 2,700 by July 1944. The peak year of production was 1944 when 23,169 engines were built, an average of 1,930 per month. Each Packard Merlin cost an average $US12,000.

Packard built numerous Merlin derivatives in both two speed/one stage and two speed/two stage supercharger forms for the P-40 Kittyhawk, Hurricane, Spitfire, Lancaster, P-51 Mustang and others. The company introduced some innovations to the Merlin: Bendix-Stromberg injection carburettors, automatic supercharger gear shifts, water-alcohol injection, a ball bearing main water pump (replacing Rolls-Royce's plain bearing unit), a centrifugal air/oil separator to prevent foaming and a new magneto for high altitudes among them.

The Packard Merlin variants produced for the Lancaster B.III were similar to the Rolls-Royce versions installed in the Lancaster I: the Merlin 28 and 38 (equivalent to the RR Merlin 22) producing 1,300hp (970kW) for takeoff and 1,240hp (925kW) at 11,500 feet, and the Merlin 224 (equivalent to the British Merlin 24 with greater boost and power) rated at 1,620hp (1,210kW) for takeoff and 1,500hp (1,120kW) at 9,500 feet.

Two Lancaster IIIs: Yeadon built LM321/PH-H (top) of 12 Squadron and Woodford built ND521/AR-F 'squared' of 460 Squadron RAAF. Note the unusual placement of the squadron code letters and serial number on PH-H. (via Neil Mackenzie)

The last Lancaster built at Avro's Yeadon (Yorkshire) facility (B.III TX273) takes off from the factory in October 1945. Yeadon was responsible for the production of 695 Lancasters from late 1942, of which all but the first 54 were B.IIIs.

Lancaster B.III Special

This designation was applied to 23 Lancaster B.IIIs modified to participate in the May 1943 Ruhr Dams raid by aircraft from 617 Squadron – Operation Chastise.

All 23 aircraft were from the Woodford built ED— serial block, powered by Packard Merlin 28s and delivered in March and April 1943. A prototype conversion to this new, 'special' standard (ED765) was flown in its modified guise in early April and used for initial trials with inert 'Upkeep' mines, the 9,250lb (4,196) Barnes Wallis designed 'bouncing bomb' used in the raid. Two others (ED817 and ED825) were also used for initial trials.

The Lancasters had their bomb doors and dorsal turret removed, the former necessary to accommodate the cylindrical weapon and the latter to save weight, while the front and rear parts of the bomb bay were faired in. The bomb was carried between two V-struts and captured by disc wheels on the struts which were mated with circular tracks on each end of the weapon. One of these wheels was hydraulically belt driven so as to spin the mine backwards to the required 500rpm before release.

Another special requirement of the mission was that the bombs be released at precisely 60 feet (18.2m) above the water onto which they would be dropped. The necessary accuracy was achieved by fitting two Aldis lamps, one in the nose and the other at the rear of the bomb bay. These two 'spotlights' pointed downwards and slightly to the side of the aircraft (so they could be seen from the cockpit) and their beams were set to converge in a figure-eight pattern when the Lancaster was at 60 feet. This idea was not derived from a moment of inspiration by the raid's leader, Wng Cdr Guy Gibson, at a London theatre (as presented in the film *The Dambusters*), but was adapted from a similar technique already tried by RAF Coastal Command for judging height when attacking U-boats at night.

After the dams raid (described in the following chapter), the surviving Lancaster B.III Specials reverted to standard specification.

Postwar Lancaster IIIs

Three new Lancaster III subvariants emerged after World War II, all of them associated with maritime activities and all conversions of existing aircraft. The first was the **Lancaster ASR.III**, developed for the air-sea-rescue role, mainly for use in the Pacific and Far East.

A prototype conversion (ND589) was carried out by Avro with Cunliffe Owen Aircraft of Eastleigh performing 120 conversions to ASR.III standards over a two year period from the second half of 1945, the Lancasters involved being taken from storage. The main feature of the ASR.III was the carriage of a droppable Mk.IIA airborne lifeboat of 30ft 6in (9.3m) length under the bomb bay. The Lancaster ASR.III served with nine RAF Coastal Command squadrons at home and abroad.

By 1948 the need for a general reconnaissance version had come about due to the return of Coastal Command's Consolidated Liberators to the USA under the terms of Lend-Lease. Thus evolved the **Lancaster GR.III** (GR.3 when Arabic replaced Roman numerals in British military aircraft designations shortly after the war), about 100 of which were modified to the new standard from the ASR.III conversions.

These retained the ability to carry a lifeboat and were fitted with ASV III (air-surface-vessel) radar in a radome under the rear fuselage. The mid upper gun turret was removed from both the ASR and GR.III, observation windows were added to the starboard rear fuselage and in the case of the GR.3, a rear facing camera was mounted under the tail turret which in both versions had its guns removed and was used as an observation point. The GR.3s were redesignated MR.3 (maritime reconnaissance) in 1950.

The RAF's last active Lancaster was GR.3 RF325 of the School of Maritime Reconnaissance at St Mawgan, Cornwall. It was retired in October 1956.

Lancaster ASR.III RF310 of 38 Squadron postwar, one of 120 aircraft converted for air-sea rescue duties from late 1945. Note the underslung Mk.IIA droppable lifeboat of 30ft 6in (9.3m) in length. RF310 was built in May 1945 by Armstrong Whitworth at Coventry.

LANCASTER PRODUCTION

Notes: This table summarises Lancaster production between 1941 and 1946 and amounts to 7,377 aircraft comprising three prototypes (two Mk.Is and one Mk.II), 3,434 Mk.Is, 300 Mk.IIs, 3,030 Mk.IIIs, 180 Mk.VIIs and 430 Canadian built Mk.Xs. RAF serial numbers are noted.

The factories noted are those at which manufacture of the aircraft took place. In some cases the aircraft were manufactured at one site and assembled at another, notably those built by Metropolitan-Vickers at Mosley Road, Manchester, most of which were assembled by Avro at Woodford (Manchester).

The manufacturers and factories involved are Avro Woodford (Manchester) and Yeadon (Yorkshire), Metropolitan-Vickers Mosley Road (Manchester), Armstrong Whitworth (Coventry), Austin Motors Longbridge (Birmingham), Vickers-Armstrong Castle Bromwich (Birmingham), Vickers-Armstrong (Chester) and Victory Aircraft (Malton, Canada).

Delivery dates are in some cases approximate but nevertheless give a good idea of when specific aircraft were built. The Canadian built aircraft were allocated serial numbers which were apparently out of order with the KB— batch built first followed by those with FM— serials.

Abbreviations: del delivery; ff- first flight; Qty – quantity.

Mark	Qty	RAF Serials	Factory	Remarks
I	1	BT308	Woodford	1st prototype ff 09/01/41
I	1	DG595	Woodford	ff 13/05/41
II	1	DT810	Woodford	first with Hercules, ff 26/11/41
I	23	L7527-7549	Woodford	del 11/41-03/42
I	20	L7565-7584	Woodford	del 01-03/42
I	36	R5482-5517	Woodford	del 02-03/42
I	40	R5537-5576	Woodford	del 03-04/42
I	38	R5603-5640	Woodford	del 04-05/42
I	46	R5658-5703	Woodford	del 05-06/42
I	40	R5724-5763	Woodford	del 06-07/42
I	27	R5842-5868	Mosley Road	del 01-05/42
I	30	R5888-5917	Mosley Road	del 05-09/42
I	39	W4102-4140	Woodford	del 07-08/42
I	48	W5154-4201	Woodford	del 08-09/42
I	50	W4230-4279	Woodford	del 09-10/42
I	40	W4301-4340	Woodford	del 10-11/42
I	30	W4355-4384	Woodford	del 11/42
I	40	W4761-4800	Mosley Road	del 09-11/42
I	50	W4815-4864	Mosley Road	del 12/42-01/43
I	27	W4879-4905	Mosley Road	del 01-02/43
I	50	W4918-4967	Mosley Road	del 02-04/43
I	3	W4980-4982	Mosley Road	del 04/43
III	30	W4983-5012	Mosley Road	del 04-05/43
II	35	DS601-635	Coventry	del 09-12/42
II	46	DS647-692	Coventry	del 12/42-03/43
II	38	DS704-741	Coventry	del 03-05/43
II	41	DS757-797	Coventry	del 05-07/43
II	40	DS813-852	Coventry	del 08-10/43
III	48	DV155-202	Mosley Road	del 05-07/43
III	30	DV217-246	Mosley Road	del 07-08/43
III	14	DV263-276	Mosley Road	del 08/43
I	6	DV277-282	Mosley Road	del 08/43
III	8	DV283-290	Mosley Road	del 08/43
I	8	DV291-297	Mosley Road	del 08-09/43
III	1	DV298	Mosley Road	del 09/43
I	14	DV299-312	Mosley Road	del 09/43
I	22	DV324-345	Mosley Road	del 09-10/43
I	49	DV359-407	Mosley Road	del 10-11/43
I	32	ED303-334	Woodford	del 11/42
I/III	50	ED347-396	Woodford	del 11-12/42
I/III	46	ED408-453	Woodford	del 12/42
I/III	38	ED467-504	Woodford	del 12/42-01/43
I/III	50	ED520-569	Woodford	del 01/43
I/III	49	ED583-631	Woodford	del 01-02/43
I/III	24	ED645-668	Woodford	del 02/43
I/III	50	ED688-737	Woodford	del 02/43
I/III	38	ED749-786	Woodford	del 02-03/43
III	44	ED799-842	Woodford	del 03/43
III	33	ED856-888	Woodford	del 03-04/43
III	50	ED904-953	Woodford	del 04/43
III	33	ED967-999	Woodford	del 05/43
III	46	EE105-150	Woodford	del 05/43
III	37	EE166-202	Woodford	del 05-06/43
X	130	FM100-229	Canada	del 03-05/45
I	45	HK535-579	Castle Brom	del 12/43-03/44
I	36	HK593-628	Castle Brom	del 03-06/44
I	21	HK644-664	Castle Brom	del 06-07/44
I	32	HK679-710	Castle Brom	del 07-09/44
I	46	HK728-773	Castle Brom	del 09-12/44
I	20	HK787-806	Castle Brom	del 01-02/45
III	47	JA672-718	Woodford	del 06/43
III	34	JA843-876	Woodford	del 06-07/43
III	50	JA892-941	Woodford	del 07/43
III	25	JA957-981	Woodford	del 07/43
III	43	JB113-155	Woodford	del 08/43
III	18	JB174-191	Woodford	del 08/43
III	28	JB216-243	Woodford	del 08-09/43
III	46	JB275-320	Woodford	del 09/43
III	33	JB344-376	Woodford	del 09-10/43
III	27	JB398-424	Woodford	del 10/43
III	36	JB453-488	Woodford	del 10/43
III	42	JB526-567	Woodford	del 10-11/43
III	23	JB592-614	Woodford	del 11/43
III	48	JB637-684	Woodford	del 11/43
III	50	JB699-748	Woodford	del 11-12/43
X	300	KB700-999	Canada	del 09/43-03/45
II	37	LL617-653	Coventry	del 10/43-12/43
II	39	LL666-704	Coventry	del 12/43-02/44
II	24	LL716-739	Coventry	del 02-03/44
I	19	LL740-758	Coventry	del 11/43
I	43	LL771-813	Coventry	del 11/43-12/43
I	42	LL826-867	Coventry	del 12/43-01/44
I	44	LL880-923	Coventry	del 01-02/44
I	43	LL935-977	Coventry	del 02-03/44
I	43	LM100-142	Coventry	del 03-05/44
I	37	LM156-192	Coventry	del 05-06/44
I	39	LM205-243	Coventry	del 06-07/44
I	40	LM257-296	Coventry	del 07-08/44
I	10	LM301-310	Yeadon	del 11-12/42
III	36	LM311-346	Yeadon	del 12/42-02/43
III	37	LM359-395	Yeadon	del 02-04/43
III	32	LM417-448	Yeadon	del 04-06/43
III	44	LM450-493	Yeadon	del 06-09/43
III	45	LM508-552	Yeadon	del 09-12/43
III	31	LM569-599	Yeadon	del 12/43-02/44
III	44	LM615-658	Yeadon	del 02-05/44
III	27	LM671-697	Yeadon	del 05-07/44
III	44	LM713-756	Yeadon	del 07-10/44
III	43	ME295-337	Yeadon	del 10-11/44
I/III	46	ME350-395	Yeadon	del 11-12/44
I/III	42	ME417-458	Yeadon	del 12/44-01/45
I/III	34	ME470-503	Yeadon	del 01-02/45
III	35	ME517-551	Yeadon	del 02-03/45
I	43	ME554-596	Mosley Road	del 11/43
I	38	ME613-650	Mosley Road	del 11-12/43
I	42	ME663-704	Mosley Road	del 12/43
I	43	ME717-759	Mosley Road	del 12/43-01/44
I	42	ME773-814	Mosley Road	del 01/44
I	42	ME827-868	Mosley Road	del 01/44
III	45	ND324-368	Woodford	del 12/43
III	46	ND380-425	Woodford	del 12/43
III	42	ND438-479	Woodford	del 12/43-01/44
III	47	ND492-538	Woodford	del 01/44
III	47	ND551-597	Woodford	del 01-02/44
III	46	ND613-658	Woodford	del 02/44
III	45	ND671-715	Woodford	del 02/44
III	42	ND727-768	Woodford	del 02-03/44
III	46	ND781-826	Woodford	del 03/44
III	44	ND839-882	Woodford	del 03/44
III	42	ND895-936	Woodford	del 03-04/44
III	49	ND948-996	Woodford	del 04/44

LANCASTER PRODUCTION (cont)

Mark	Qty	RAF Serials	Factory	Remarks
III	40	NE112-151	Woodford	del 04-05/44
III	19	NE163-181	Woodford	del 05/44
I	34	NF906-939	Coventry	del 07/44
I	48	NF952-999	Coventry	del 07-08/44
I	37	NG113-149	Coventry	del 08/44
I	45	NG162-206	Coventry	del 08-09/44
I	42	NG218-259	Coventry	del 09-10/44
I	46	NG263-308	Coventry	del 10/11-44
I	47	NG321-367	Coventry	del 11-12/44
I	43	NG379-421	Coventry	del 12/44-01/45
I	36	NG434-469	Coventry	del 01/45
I	22	NG482-503	Coventry	del 01-02/45
I	33	NN694-726	Longbridge	del 03-07/44
I	48	NN739-786	Longbridge	del 07/44-01/45
I	19	NN798-816	Longbridge	del 01-02/45
I	42	NX548-589	Longbridge	del 02-04/45
I	8	NX603-610	Longbridge	del 04/45
VII	38	NX611-648	Longbridge	del 04-05/45
VII	43	NX661-703	Longbridge	del 05-07/45
VII	44	NX715-758	Longbridge	del 07-08/45
VII	25	NX770-794	Longbridge	del 08-09/45
I	41	PA158-198	Chester	del 06-09/44
I	26	PA214-239	Chester	del 09-10/44
I	37	PA252-288	Chester	del 10-12/44
I	49	PA303-351	Chester	del 12/44-03/45
I	32	PA365-396	Chester	del 03-05/45
I	43	PA410-452	Chester	del 06-09/45
I	6	PA473-478	Chester	del 09/45
I	1	PA509	Chester	del 09/45
III	36	PA964-999	Woodford	del 05/44
III	47	PB112-158	Woodford	del 05/44
III	43	PB171-213	Woodford	del 05-06/44
III	42	PB226-267	Woodford	del 06/44
III	29	PB280-308	Woodford	del 06/44
III	45	PB341-385	Woodford	del 06-07/44
III	42	PB397-438	Woodford	del 07/44
III	41	PB450-490	Woodford	del 07-08/44
III	39	PB504-542	Woodford	del 08/44
III	43	PB554-596	Woodford	del 08/44, PB592 was Mk.I
III	34	PB609-642	Woodford	del 08-09/44
I/III	11	PB643-653	Woodford	del 10/44
I/III	43	PB666-708	Woodford	del 10-11/44
I/III	48	PB721-768	Woodford	del 11/44
I	44	PB780-823	Woodford	del 11-12/44
I	46	PB836-881	Woodford	del 12/44-01/45
I	44	PB893-936	Woodford	del 01-02/45, PB 923 was Mk.III
I/III	46	PB949-994	Woodford	del 02-03/45
I(Sp)	4	PB995-998	Woodford	del 01/45, to carry Grand Slam bomb
I(Sp)	28	PD112-139	Woodford	del 01-03/45, for Grand Slam bomb

Mark	Qty	RAF Serials	Factory	Remarks
I	42	PD198-239	Mosley Road	del 06-07/44
I	45	PD252-296	Mosley Road	del 07-09/44
I	41	PD309-349	Mosley Road	del 09-10/44
I	44	PD361-404	Mosley Road	del 10-11/44
I	28	PD417-444	Mosley Road	del 11-12/44
I	33	PP663-695	Castle Brom	del 02-04/45
I	46	PP713-758	Castle Brom	del 04-07/45
I	21	PP772-792	Castle Brom	del 07-08/45
I	48	RA500-547	Mosley Road	del 12/44-02/45
I	48	RA560-607	Mosley Road	del 02-05/45
I	5	RA623-627	Mosley Road	del 05/45
I	20	RA787-806	Mosley Road	del 05-06/45
III	26	RE115-140	Yeadon	del 03-04/45
III	36	RE153-188	Yeadon	del 04-05/45
III	23	RE200-222	Yeadon	del 05-06/45
III	2	RE225-226	Yeadon	del 06/45
I	42	RF120-161	Coventry	del 02-03/45
I	23	RF175-197	Coventry	del 03/45
III	19	RF198-216	Coventry	del 03/45
III	45	RF229-273	Coventry	del 04-05/45
III	41	RF286-326	Coventry	del 05/45
VII	30	RT670-699	Longbridge	del 11/45-01/46
I	37	SW243-279	Mosley Road	del 11-12/44
III	13	SW283-295	Coventry	del 05-06/45
I	21	SW296-316	Coventry	del 06/45
III	27	SW319-345	Yeadon	del 06-08/45
III	20	SW358-377	Yeadon	del 08-09/45
I(FE)	25	TW647-671	Coventry	del 06-07/45
I	16	TW858-873	Coventry	del 07-10/45
I	34	TW878-911	Coventry	del 10/45-03/46, final batch
I	15	TW915-929	Chester	del 06-08/45
III	11	TX263-273	Yeadon	del 09-10/45

LANCASTER PRODUCTION BY SITE

	Proto	Mk.I	Mk.II	Mk.III	Mk.VII	Mk.X	Totals
Avro (Woodford)	3	840	–	2135	–	–	2978
Avro (Yeadon)	–	54	–	641	–	–	695
Metrovick (Woodford)	–	921	–	136	–	–	1057
AWA (Coventry)	–	911	300	118	–	–	1329
Austin (Birmingham)	–	150	–	–	180	–	330
Vickers (Castle Brom)	–	300	–	–	–	–	300
Vickers (Chester)	–	258	–	–	–	–	258
Victory (Canada)	–	–	–	–	–	430	430
Totals	**3**	**3434**	**300**	**3030**	**180**	**430**	**7377**

Note: The figures reflect the factory at which final assembly took place. In some cases the aircraft was manufactured at one site and assembled elsewhere. Most Lancasters manufactured by Metropolitan-Vickers at Mosley Road Manchester were completed at Avro's Woodford (Manchester) facility.

LANCASTER IV and V

Designations given to planned improved Lancasters powered by more powerful Rolls-Royce Merlin 85s or Packard Merlin 68s with two speed/two stage superchargers. Developed as the Avro Type 694 Lincoln B.1 and B.2, respectively, as described later. The first aircraft was flown (as the Lancaster IV) on 9 June 1944 and redesignated Lincoln the following August.

LANCASTER VI

Rolls-Royce's ongoing development of the Merlin series resulted in numerous variants, the company developing a family of engines with two speed/two stage superchargers from 1941. Among this group was the Merlin 85 rated at 1,635hp (1,220kW) for takeoff and 1,750hp (1,305kW) at 5,500 feet. Featuring annular radiators enclosed by circular section engine cowlings, the Merlin 85 was developed in 1943 for what was originally named the Lancaster IV but eventually emerged as the Lincoln B.1.

Two Lancaster IIIs were assigned to Rolls-Royce in mid 1943 for use as test beds for this engine, while a third Lancaster III (JB675) was flown with Merlin 85s in January 1944. Despite being the third of its type to fly, this aircraft is regarded as being the prototype for what became known as the Lancaster VI.

Nine Lancasters were converted to Mk.VI configuration, four of which flew temporarily with several squadrons including on limited operations with Nos 7 and 635 Squadrons between August and November 1944, logging about 20 sorties. They were mainly used for radar jamming and for creating diversions. The extra performance offered by the more power-

Still in French Navy markings, Lancaster VII G-ASXX (the former RAF NX611) is escorted by an RAAF Canberra and RAF Victor in the early stages of its flight from Australia to Britain in 1965. The aircraft is now preserved in ground running order (in RAF colours) by the Lincolnshire Aviation Heritage Centre.

ful Merlin 85 was useful but was balanced by a lack of reliability as the engine was still at an early stage of its development.

The full expression of the Merlin 85's usefulness would come a bit later in the form of the Lincoln.

LANCASTER B.VII

The Lancaster VII was basically a Lancaster I with Merlin 24s fitted with an electrically operated Martin 250 CE 23A dorsal turret containing two 0.50in machine guns instead of the usual hydraulically driven Frazer Nash FN 50 with two 0.303s. It was produced exclusively by Austin Motors at Longbridge, Birmingham. Production amounted to 180 aircraft, delivered between April 1945 and January 1946.

The new turret was mounted further forward than its predecessor to a point just aft of the wing's trailing edge. This position improved the gunner's ease of entry to the turret – and exit in an emergency – but somewhat restricted movement around the aircraft's fuselage. The Mk.VII was the heaviest of the standard production Lancasters with a maximum weight of 70,000lb (31,752kg), compensated for to some extent with the fitting of 1,620hp (1,210kW) Merlin 24s.

Austin also built 150 Lancaster Is from early 1944, 50 of which were intended to be Mk.VIIs but due to a shortage of Martin turrets had FN 50s fitted in the new forward position. First trials of the Martin turret were carried out in a Lancaster III from March 1944.

Intended to operate as part of the planned 'Tiger Force' in the Far East on long range missions against Japan (an idea abruptly terminated with the atomic bomb and the end of the Pacific war), no Lancaster VIIs saw

wartime service, although the variant did serve with several squadrons in the Far East postwar. The incorporation of a tropicalisation kit resulted in the designation Lancaster B.VII (FE). One of the modifications incorporated was the fitting of a new rear turret, the Frazer Nash FN 82 with a pair of 0.50in machine guns installed.

LANCASTER B.X

North American industry not only provided support to the Lancaster programme in the form of Packard Merlin engines from the USA, but also by way of 430 complete aircraft

manufactured by Canada's Victory Aircraft between 1943 and 1945.

Basically similar to the Lancaster III with Packard Merlins and minor equipment differences, the Canadian version was designated B Mk.X and stemmed from early 1942 talks between the British and Canadian governments with a view to supplementing British Lancaster production. Agreement was reached and Victory Aircraft established in order to manufacture Lancasters under licence. Owned by the Canadian Government but established in conjunction with the National Steel Corporation of Canada,

Two Canadian built Lancaster B.Xs: KB700 (top, the first aircraft) with Frazer Nash mid upper turret and KB783 (bottom) with Martin turret and 0.50in machine guns. Both have the bulged bomb bay, a feature of most Canadian Lancasters. Victory aircraft built 430 Lancasters in 1943-45.

Postwar conversions resulted in several new Lancaster 10 subvariants for service with the Royal Canadian Air Force. Illustrated is a Lancaster 10-N for navigation training.

Victory was set up in the Malton (Ontario) plant of the National Steel Car Corporation. Postwar, the company was purchased by Hawker Siddeley and became known as Avro Canada.

A 'pattern' Lancaster I (R5727) was sent to Canada in August 1942 and while work preparations for production were underway, orders were placed for 500 aircraft in two batches. Seventy of these were cancelled with the end of the war.

Just under a year after the arrival of the pattern aircraft, the first Canadian Lancaster B.X (KB700) was flown on 6 August 1943. Deliveries began the following September, all but the last 22 aircraft finding their way across the Atlantic for service with the RAF, mainly with the squadrons comprising the Canadian No 6 Group in Britain. Canadian production gradually accelerated, the 100th aircraft coming off the line in August 1944 and the 300th in March 1945. By that stage the production rate was one per day and the last example was delivered in May 1945.

The first 75 Lancaster Xs were powered by 1,300hp (970kW) (for takeoff) Packard Merlin 38s, while the remainder had 1,620hp (1,210kW) Packard Merlin 224s. Most were fitted with the bulged bomb bay capable of accommodating an 8,000lb (3,270kg) weapon, while a ventral turret with two 0.50in machine guns was also a usual fit. Later aircraft had their Frazer Nash mid-upper turrets replaced with the Martin unit fitted to the British built Lancaster B.VII.

Postwar Tens

The postwar Royal Canadian Air Force had 228 Lancaster Xs on strength, comprising some aircraft which had not crossed the Atlantic to join the RAF and about 200 survivors of the war which returned to their country of origin after hostilities had ended.

Many of these were subsequently modified to fill other roles, resulting in some new designations – *Lancaster 10-BR:* (bomber-reconnaissance) 13 conversions; *10-MR:* (maritime reconnaissance) 72 conversions of which most were eventually redesignated *10-MP* (maritime patrol); *10-SR:* (air-sea-rescue) eight conversions; *10-P:* (photo reconnaissance) 11 conversions; *10-N:* (navigation trainer) three conversions; and *10-D:* (drone carrier) two conversions.

The RCAF retired its last Lancaster in April 1964.

TRANSPORT LANCASTERS

The first Lancaster transport conversion was made to Lancaster I R5727, the pattern aircraft supplied to Victory Aircraft as a prelude to production of the aircraft in Canada. Following its use by the manufacturer, the Lancaster was acquired by Trans-Canada Air Lines (TCA) and used for experimental freight flights from early 1943. The aircraft had its turrets removed and faired over and windows were fitted to the rear fuselage. It was given the Canadian civil registration CF-CMS.

It was then further modified by Avro to incorporate extra fuel tanks and to carry 10 passengers, in this form inaugurating a trans-Atlantic service between Dorval (Montreal) and Prestwick from July 1943 carrying forces mail initially and then priority freight and passengers as well. It was later joined by eight similar conversions in 1944-45, the last six of which featured lengthened noses in which mail was carried, new faired tail sections, two 400 Imp gal (1,818 l) fuel tanks in the bomb bay giving a range of up to 4,150 miles (6,678km), lined cabins, ceiling lights and in the case of the last four aircraft, an attempt at soundproofing the cabin. The original Packard Merlin 38s were eventually replaced by more powerful 224s.

The later conversions were dubbed Lancaster XPPs (for 'Mk.X Passenger Plane') and paved the way for the Avro produced Lancastrian. They remained in trans-Atlantic service until 1947 – carrying revenue passengers after the end of the war in Europe – when they were replaced by Canadair Four North Stars, in the meantime extending their service to Heathrow.

The first Lancaster transport conversion, from B.I R5727, the aircraft provided to Victory Aircraft as a pattern aircraft prior to Canadian production. As Trans-Canada Air Lines' CF-CMS, the Lancaster inaugurated a trans-Atlantic service in July 1943. (BAe via Philip J Birtles)

Another view of CF-CMS, photographed at Prestwick in 1943.

The first British Lancaster transport conversion, Metropolitan-Vickers built B.I DV379/G-AGJI. The conversion was performed in late 1943/early 1944.

The original conversion (R5727/CF-CMS) was eventually fitted with Merlin 85 engines as installed in the Lancaster VI (and Lincoln) but was lost in a takeoff accident at Dorval in June 1945.

The Lancastrian

The success of the Lancaster XPP project resulted in Avro deciding to produce its own Lancaster transport under the name Type 691 Lancastrian. The winding down of Lancaster production in 1945 allowed airframes on the production line to be completed as Lancastrians, a total of 82 eventually being built at Woodford. The first (PD180, later VB873/G-AGLF) was flown on 17 January 1945 and like other very early aircraft had passenger windows only on the starboard side of the fuselage. Later Lancastrians had windows on both sides.

The basic aircraft were externally similar to the XPP with military equipment removed and lengthened fairings covering the nose and tail sections. Internally, most featured more luxurious interiors with nine seats initially and 13 later. Although a very basic passenger transport with a cramped and unpressurised cabin, the Lancastrian found itself in service for rather longer than it should have been due to the failure of Avro's first purpose built postwar airliner, the Tudor.

The first batch of 23 Lancastrian Mk.1s were delivered to BOAC during 1945 for use on what was later called the 'Kangaroo Route' between England and Australia. With accommodation for only nine passengers (or six at night) plus freight and mail,

these aircraft were hardly economic propositions but they were relatively fast and had good range. Besides, there was nothing else available! The Lancastrian 1 was powered by four 1,620hp (1,210kW) Merlin T.24s and in addition to the 21 aircraft for BOAC, another two were built for the RAF as VH737 and VH742. These were used as testbeds for the Rolls-Royce Nene jet engine.

Most Lancastrian 1s had been scrapped by 1950 although two were purchased by Skyways and three by Qantas.

The Lancastrian C.2 was the RAF's equivalent of the civil Mk.1; 33 of them were built with the serial numbers VL967-981, VM701-704 and VM725-738. The first example was delivered in October 1945.

In November 1945 a 511 Squadron Lancastrian achieved some fame by flying around the world, covering 36,000 miles (57,900km) in 36 days. This was the first of many long distance flights by Lancastrians, an even more notable one being recorded by VM726 in March 1946 when it became the first aircraft to circumnavigate the globe in less than a week.

RAF Lancastrians were used for general transport duties, conversion training and by the Empire Flying School and Empire Air Navigation School. Seven became engine testbeds, this very important role seeing just about every new British engine of the late 1940s/early 1950s being installed on a Lancastrian, Lancaster or Lincoln. The Lancastrian testbeds were VH742 and VH737 (RR Nenes in the outer positions, first flight August 1946); VM703 and VM729 (DH Ghosts, first flight July 1947); VM732 and VL970 (RR Avons); and VM733 (Armstrong Siddeley Sapphires). VM704 and VM728 had RR Griffon piston engines on the inner positions.

Following RAF service, some Lancastrian C.2s were sold to BOAC, British South American Airways (BSAA) and Skyways.

The prototype Avro Lancastrian, G-AGLF. Compare the nose and tail fairings with those on earlier Lancaster transports.

Qantas Lancastrian 1 VH-EAU, the former BOAC G-AGLZ. (via Bob Livingstone)

The Lancastrian 3 was developed for BSAA as a shorter range 13 seat airliner, and although 18 were ordered only six were delivered. The remainder went to Alitalia, Silver City Airways and Qantas.

The final Lancastrian variant was the RAF's C.4, eight of which (TX283-290) were ordered in late 1945 for delivery in mid 1946. These aircraft were equipped as 13 seaters but saw very limited – if any – RAF service before they were refurbished and sold to Skyways and *Flota Aerea Mercante Argentina*. The Skyways aircraft were extensively used in the 1948 Berlin Airlift, as were aircraft belonging to Flight Refueling Ltd, which had a mixed fleet of second hand aircraft. These were converted to tankers for the purpose of delivering bulk fuel to beleaguered Berlin, and had 2,500imp gal (11,365 l) fuselage tanks installed.

> ### LANCASTRIAN Mk.1
> **Powerplants:** *Four Rolls-Royce Merlin T.24 liquid cooled V12 piston engines rated at 1,620hp (1,210kW) for takeoff and 1,500hp (1,120kW) at 9,500ft; De Havilland constant-speed and feathering three bladed propellers of 12ft 0in (3.66m) diameter; fuel capacity 3,174imp gal (14,429 l).*
> **Dimensions:** *Wing span 102ft 0in (31.09m); length 76ft 10in (23.42m); height 19ft 6in (5.94m); wing area 1,297sq ft (120.5m²).*
> **Accommodation:** *Nine passengers in cabin plus space for light freight and mail.*
> **Weights:** *Empty 30,426lb (13,801kg); max loaded 65,000lb (29,484kg).*
> **Performance:** *Max speed 274kt (507km/h) at 12,000ft; max cruise 252kt (466km/h) at 17,500ft; economical cruise 200kt (370km/h) at 15,000ft; initial climb 950ft (290m)/min; service ceiling 25,500ft (7,772m); max range with nine passengers and 210lb (95kg) mail 3,610nm (6,678km).*

YORK

Mention should also be made of the Avro 685 York, a four engined long range transport developed in a period of just six months in 1942. Designer Roy Chadwick combined the Lancaster's wings, tail unit (with a third central fin), powerplants and undercarriage with a new, more capacious slab sided fuselage. The wings were shoulder mounted rather than in the mid position of the Lancaster.

The first York (LV626) was flown on 5 July 1942 and 257 were eventually built (including one by Victory Aircraft in Canada) for the RAF and BOAC while other civilian users postwar, flew mainly ex RAF aircraft. All but one of the Yorks were Mk.Is with Rolls-Royce Merlin engines; the sole Mk.II was powered by Bristol Hercules radials.

The last RAF York was disposed of in 1957 while the last civilian examples were retired in the early 1960s.

The Avro York transport, combining the Lancaster's wings, engines, tail unit (with an added central fin) and undercarriage with a capacious new fuselage. The first York was flown in July 1942 and 257 were eventually built. (DoD)

LANCASTER TESTBEDS

While the Lancaster was used extensively during the war years to test new weapons and items of equipment, it also made a significant contribution to the development of new aero engines in the postwar era. All the major British engine manufacturers used Lancasters as testbeds for development of their jet engines with the exception of Rolls-Royce, which used Lancastrians and Lincolns. Rolls-Royce did however use a Lancaster to test its Dart turboprop.

The following table lists the Lancasters used as engine testbeds in Britain and abroad and for the sake of completeness, the Lancastrians and Lincolns are included:

LINCOLN

The story of the Avro Lincoln is in many ways that of the poor relation. Here was a bomber with impeccable breeding but due to its timing gained none of the fame and legendary status of its immediate and closely related predecessor, the Lancaster. Intended as a higher flying, longer ranging derivative of its famous forebear, the Lincoln was designed as a Lancaster replacement initially for use in Europe and then to equip the Royal Air Force in the Pacific War against Japan.

Delays in changing over from Lancaster to Lincoln production in late 1944 and early 1945 prevented the Lincoln from entering RAF squadron service until late August 1945 – and then only for service trials – by which time the European war against Germany had been over for three months and the Pacific war had just come to an abrupt end following the dropping of atomic bombs on Hiroshima and Nagasaki. The Lincoln finally entered regular service in February 1946.

Ostensibly, the Avro Type 694 Lincoln was simply a longer, heavier, better armed, more powerful and increased wing span development of the Lancaster, the original intention being to use as many Lancaster parts as possible. What would become the Lincoln was originally dubbed the Lancaster Mk.IV but the extent of the redesign was such that a new type number and name was justified.

Like all military projects of the time, the Lincoln suffered massive cutbacks in the immediate postwar period but 550 of them in two basic versions came out of the English factories of Avro, Metropolitan Vickers and Armstrong Whitworth Aircraft between the first flight of the prototype on 9 June 1944 and April 1951. To this tally must be added a single example built by Canada's Victory Aircraft in 1945 (after the company had produced 430 Lancasters) and 73

Testbeds all (top to bottom): Lancaster III SW342 with Mamba installation in nose and Adder/Viper in tail; Lancastrian VM733 with Sapphires in outer nacelles; Lancastrian VM703 with Ghosts; and Lincoln RF533 used for high altitude meteorological research and as a flight laboratory.

Lancasters, Lancastrians and Lincolns found widespread use as engine testbeds. This one has a pair of DH Ghost turbojets on the outer positions. (BAe)

ENGINE TESTBEDS

Lancasters
BT308	Metrovick F2/1 in tail 1943-44.
LL735	Metrovick F2/4 Beryl in tail 1945-48.
ND784	Armstrong Siddeley ASX in nose 1943-46.
ND784	Armstrong Siddeley Mamba in nose 1947.
NG465	Rolls-Royce Dart in nose 1947-54.
RE137	Armstrong Siddeley Python mockup 1948.
SW342	Armstrong Siddeley Mamba in nose, Adder/Viper in tail 1949-56.
TW911	Armstrong Siddeley Python, two in outer nacelles 1949-53.
RA805	Swedish STAL Dovern in underbelly pod 1951-56 (serial no 8001).
FM205	Avro Canada Chinook, two in outer nacelles 1951.
FM209	Avro Canada Orenda, two in outer nacelles 1950-56.

Lancastrians
VH737	Rolls-Royce Nene, two in outer nacelles 1947-55.
VH742	Rolls-Royce Nene, two in outer nacelles 1946.
VL970	Rolls-Royce Avon, two in outer nacelles 1947-55.
VM703	De Havilland Ghost, two in outer nacelles 1947.
VM729	De Havilland Ghost, two in outer nacelles.
VM732	Rolls-Royce Avon, two in outer nacelles 1947-50.
VM733	Armstrong Siddeley Sapphire, two in outer nacelles 1950-54.

Lincolns
RA643	Bristol Phoebus in bomb bay 1947.
RA716	Bristol Theseus, two in outer nacelles 1947.
RA716	Rolls-Royce Avon, two in outer nacelles 1955-57.
RE339	Armstrong Siddeley Python, two in outer nacelles.
RE339	Bristol Theseus, two in outer nacelles.
RE418	Bristol Theseus, two in outer nacelles 1948.
RF368	Bristol Proteus, two in outer nacelles.
RF402	Napier Naiad (mockup) in nose 1948.
RF403	Armstrong Siddeley Python, two in outer nacelles.
RF530	Napier Naiad in nose 1946.
RF533	Rolls-Royce Tyne in nose 1956.
SX971	Rolls-Royce Derwent in underbelly pod 1950-56.
SX972	Bristol Proteus, two in outer nacelles 1950.
SX973	Napier Nomad in nose 1953.

manufactured under licence in Australia by the Government Aircraft Factory (GAF) and delivered to the Royal Australian Air Force between 1946 and 1953.

A Temporary Expedient

In RAF service the Lincoln became very much a temporary expedient pending the arrival of the jet bombers and was withdrawn from front line service in 1955 as the RAF's last piston engined bomber. Despite its relatively brief life at the sharp end of Royal Air Force Bomber Command (serving with some 30 operational squadrons regardless), it saw action in two theatres of operation, firstly against communist terrorists in Malaya from 1950 (in conjunction with RAAF Lincolns) and in 1953 against the Mau Mau in Kenya. Following its withdrawal from front line service the Lincoln served in various secondary roles until 1963. Two such roles were airborne electronic surveillance in its very early days and the task of probing Soviet radar defences to see how effective they were. It was while performing one of these missions in March 1953 that an RAF Central Gunnery School Lincoln was shot down by a Soviet MiG-15 after it had strayed into East German airspace.

(left) The Lancaster and its close relatives played a major part in the development of air-to-air refuelling. Here, a Lancaster is linked with a Meteor F.3.

Australian built Lincoln Mk.30 A73-6. The Government Aircraft Factory built 73 Lincolns for the RAAF.

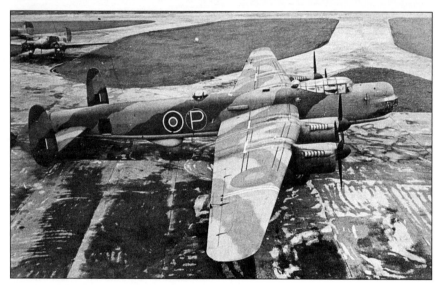

(above) The Lincoln (or Lancaster IV) prototype (PW925) at Manchester's Ringway Airport in 1944. Note the three bladed propellers.

(below) Lincoln B.2 RE290 in Far East markings. British production of the Lancaster's successor amounted to 547 aircraft.

A major part of the Lincoln's (and Lancaster's) British service was performing the role of engine test bed. As suggested earlier, the Lincoln, Lancaster and Lancastrian between them tested just about every medium or large turboprop or jet engine built in Britain in the 1940s and 1950s.

Apart from Australia, the only export customer for the Lincoln was Argentina, which took 12 refurbished ex-RAF and 18 new aircraft from 1947. For Australia, the Lincoln was an important aeroplane in that it provided the country's aircraft manufacturing industry with invaluable work in the postwar era of severe cutbacks for orders of military equipment generally. It also gave the RAAF a bomber which although obsolescent early in its career served usefully before the arrival of the jet powered Canberra. Of interest is the fact that Lincoln was also the largest aircraft ever to be built in Australia.

The RAAF found many roles for the Lincoln other than the basic one of bomber including general and maritime reconnaissance (the latter pending the arrival of purpose built Lockheed Neptunes), navigation and photographic training, anti-submarine warfare training and the subsequently controversial air sampling during British atomic bomb trials at Monte Bello Island and Maralinga.

The Royal Australian Air Force retired its last of its Lincolns in 1961.

A Better Lancaster

Considering the immense success of the Lancaster, it was only logical that by 1943 Roy Chadwick and his team should start to examine ways of producing a better performing Lancaster which could take advantage of lessons learned so far and also the availability of new Rolls-Royce Merlin

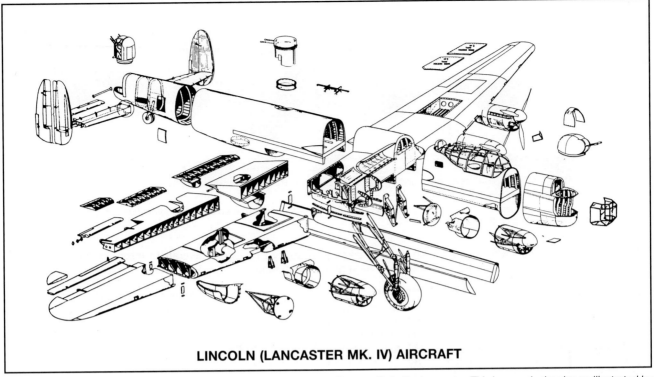

LINCOLN (LANCASTER MK. IV) AIRCRAFT

Exploded view of the Lincoln's major components, which generally applies equally to the Lancaster. This is an early drawing as illustrated by the three (instead of four) bladed propellers, the four gun rear turret and use of the words 'Lancaster Mk.IV'.

variants with two speed/two stage superchargers which promised improved altitude performance and greater bomb loads.

From these studies came the issue of Specification B.14/43 and the ordering of prototypes of the Lancaster Mks.IV and V, the former intended for aircraft with British built Merlins and the latter with Packard Merlins.

The changes incorporated in the Lancaster Mks.IV and V were such that a new type number and name were justified, the new bomber therefore becoming the Avro Type 694 Lincoln. Compared to the Lancaster it featured a wing of greater span (now 120ft 0in/36.57m), increased fuel capacity of 2,850imp gals (12,956 litres) in the wings with provision for two 400imp gal (1,818 litres) overload tanks in the bomb bay, and a rear fuselage lengthened by 8ft 11.5in (2.73m) to provide more space and to help balance the heavier two speed/ two stage Merlins.

The Lincoln B.I was fitted with British built Merlin 85s rated at 1,635hp (1,220kW) for takeoff and 1,750hp (1,305kW) at 5,500 feet and the B.II with similarly rated Packard Merlin 68s. The engines were housed in complete 'power egg' form (as were the Lancaster's) but were of a new design featuring semi-annular radiators rather than the underslung units incorporated in the Lancaster. The prototype and early production Lincolns had three bladed propellers but four bladed units were fitted early in the aircraft's career to help cure a

severe vibration problem resulting from the Merlin 85's tendency to 'hunt' and set off a sympathetic vibration with the airframe, which was perhaps exaggerated by the longer and more flexible wing. The four bladed propellers solved the problem although the reliability of that particular variety of Merlin continued to cause concern.

The Lincoln's initial maximum take-off weight was set at 75,000lb (34,020kg) – about ten per cent heavier

than the Lancaster – and gradually crept up to 82,000lb (37,200kg), a factor which tended to nullify the advantages bestowed by the uprated Merlins when it came to climb and ceiling performance. At the higher weight the Lincoln was in fact inferior to its predecessor in that regard although it was 10 to 15 knots (18 to 27km/h) faster in maximum and cruise speeds.

The Lincoln's normal maximum bomb load remained as per the Lan-

The unique to Australia 'long nose' Lincoln Mk.31 developed for maritime duties. Twenty Mk.30s were converted to this standard.

Let's hope the photographer was short – er, vertically challenged! An obviously lightly loaded Lincoln demonstrates its abilities with three of its four Merlins feathered. (via Lindon Griffith)

caster at 14,000lb (6,350kg) but its defensive armament was substantially upgraded. Several gun and turret combinations were tested but the definitive arrangement in the B.II had twin 0.50in Browning machine guns in the nose controlled remotely by the bomb aimer from his station immediately below the guns and behind redesigned, optically flat panels. The mid upper turret had a pair of 20mm Hispano cannon and the rear two 0.50 Brownings. The Lincoln B.I differed in having a pair of Brownings in all three turrets and in service with the RAF both versions often had their mid upper turrets removed.

The Lincoln carried on the good payload/range characteristics established by the Lancaster and improved on them thanks to an ability to carry more fuel for a given bomb load.

Into Production

The prototype Lincoln made its first flight from Ringway on June 9, 1944 (like its predecessors in the hands of Captain H A Brown) and the type entered operational service with 44 Squadron RAF in February 1946 after service trials with 57 Squadron had been conducted from August 1945. The Lincoln was never used for the long range bombing missions for which it was intended into Eastern Europe and the Far East with Tiger Force.

A reluctance to change Lancaster production over to the Lincoln in late 1944/early 1945 caused the newer aircraft to miss out on taking part in World War II's closing stages. That reluctance was well founded as the Lancaster was proving to be perfectly adequate for the task at hand and production of it in very large numbers was by then well established and running smoothly.

With the war obviously going the way of the Allies, the original plan was to reduce Lancaster production from 260 per month in November 1944 to 124 per month in June 1945 with production ending by November 1945. To fill a projected requirement for 2,254 aircraft, the Lincoln was intended to be phased into production early in 1945, reaching 66 per month by March and peaking at 200 per month during August.

Even the initial rate was never achieved and some indication of the low priority afforded the Lincoln is given by the fact that the second prototype did not fly until November 1944. Obviously assuming peace was not far over the horizon and with an eye on post war activities, Avro was spending more time developing transport derivatives of its bombers, such as the York and the Tudor, the latter combining the wings and powerplants of the Lincoln with a new pressurised fuselage and conventional single fin tail surfaces.

Three prototype Lincolns were built along with 82 B.Is, 465 B.IIs (B.1 and B.2 postwar), the single Canadian Mk.XV and 73 Mk.30s manufactured in Australia, the latter total including 20 converted to the uniquely Australian 'long nose' configuration as the Mk.31.

(left) The second production Lincoln B.2 (RA657) conducts refuelling trials with a Meteor F.3 in 1949. The aircraft was operating for Flight Refuelling Ltd, the company responsible for developing a practical system.

LINCOLN B.2

Powerplants: *Four Packard Merlin 68A liquid cooled V12 piston engines with two speed/two stage superchargers each rated at 1,635hp (1,220kW) for takeoff and 1,750hp (1,305kW) at 5,500ft; four bladed De Havilland constant speed and feathering propellers; fuel capacity 2,850imp gal (12,956 l) in wings, provision for one or two 400imp gal (1,818 l) tanks in bomb bay.*
Dimensions: *Wing span 120ft 0in (36.57m); length 78ft 3in (23.85m); height (tail up) 20ft 6in (6.25m); wing area 1,421sq ft (132.0m²).*
Weights: *Empty 44,188lb (20,043kg); normal loaded 82,000lb (37,195kg).*
Armament: *Two 0.50in machine guns each in nose and tail turrets, two 20mm cannon in mid upper turret; normal maximum bomb load 14,000lb (6,350kg).*
Performance: *Max speed 265kt (491km/h) at 19,000ft; cruising speed 212kt (392km/h) at 22,500ft; range cruise 187kt (346km/h) at 20,000ft; initial climb 800ft (244m)/min; service ceiling 22,100ft (6,736m); range (14,000lb load) 1,280nm (2,370km), maximum range (3,000lb load) 4,000nm (7,160km).*

The final expression of the line which began with the Manchester in 1939 – the Shackleton maritime patrol aircraft which first flew in 1949 and was retired from RAF service in that role in 1967. Incredibly, a handful flew on for no fewer than 24 years after that as early warning aircraft before finally being replaced in 1991. Illustrated are: the second prototype VW135 (top) with tailwheel undercarriage and a South African Air Force MR.3 (bottom) with tricycle undercarriage.

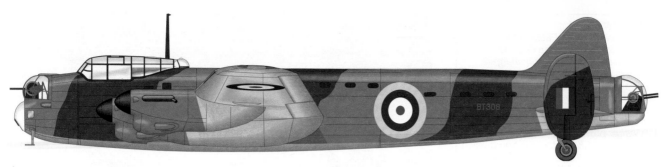

Lancaster prototype BT308 (converted on production line from Manchester airframe) at the time of its first flight on 9 January 1941.

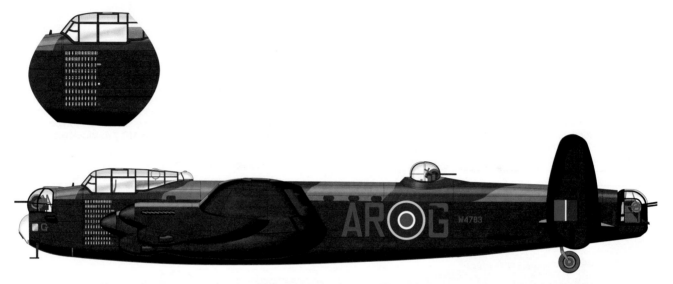

Lancaster B.I W4783/AR-G of 460 Squadron RAAF with 90 mission symbols and now preserved in the Australian War Memorial, Canberra.

Lancaster B.I R5556/KM-C of 44 (Rhodesian) Squadron RAF, the first operational unit equipped with the type in late 1941. Note the ventral gun.

Lancaster B.1 NG128/SR-B of 101 Squadron RAF, Ludford Magna 1944 with 'Cigar' aerials (ABC jamming equipment).

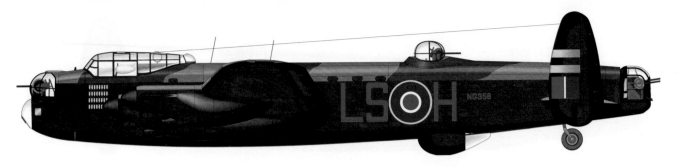

Lancaster B.I NG358/LS-H of 15 Squadron RAF, Mildenhall 1945, fitted with H2S radar. Yellow fin stripes denote Gee-H equipped formation leader.

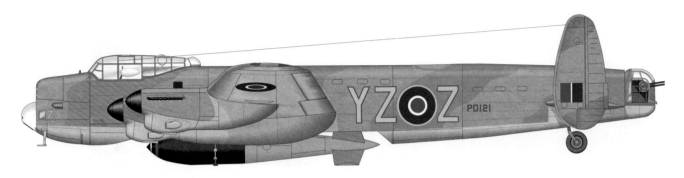

Lancaster B.I (Special) PD121/YZ-Z of 617 RAF 1945 with 22,000lb Grand Slam bomb.

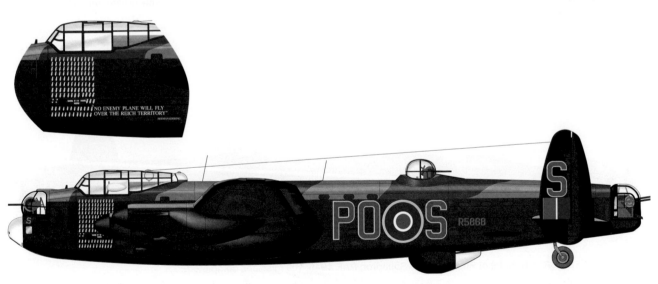

Lancaster B.I R5868/PO-S of 467 Squadron RAAF with 137 mission symbols. Now preserved in RAF Museum, Hendon.

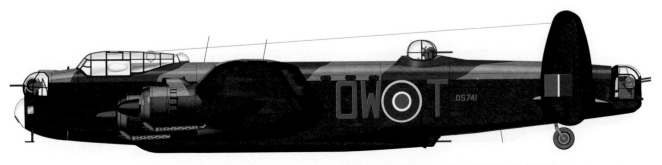

Lancaster B.II DS741/OW-T of 426 Squadron RCAF, Linton-on-Ouse 1943, with bulged bomb bay and ventral gun.

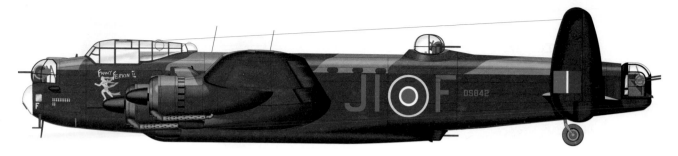

Lancaster B.II DS842/JI-F 'Fanny Ferkin II' of 514 Squadron RAF, 1944.

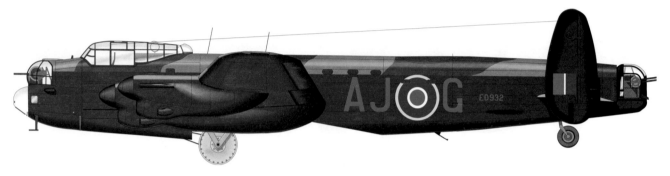

Lancaster B.III (Special) ED932/AJ-G of 617 Squadron RAF with 'Upkeep' weapon. Aircraft of Wng Cdr Guy Gibson on Dams raid, 16-17 May 1943.

Lancaster B.III EE136/CA-R 'Spirit of Russia' of 189 Squadron RAF, 1945; 109 missions completed.

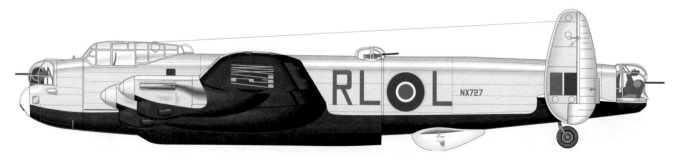

Lancaster B.VII NX727/RL-L of 38 Squadron RAF, Malta c.1947.

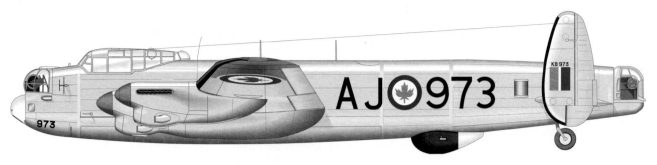

Lancaster Mk.10-MR KB973/AJ-973 of 407 Squadron RCAF, Canada 1956.

Lancaster GR.3 RE158/BS-B of 120 Squadron RAF, Leuchars 1947.

Lancaster B.I WU40 (ex RAF PA342) of Flottille 24F, Aeronavale 1953.

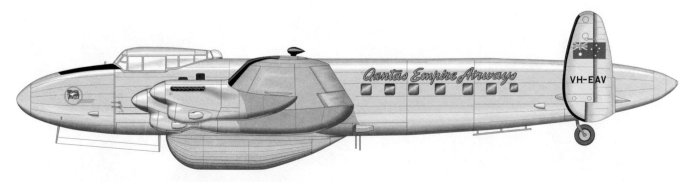

Lancastrian 3 VH-EAV of Qantas c.1950, the former Silver City Airways G-AHBW. The large underfuselage pod was capable of carrying a spare engine for Qantas' Lockheed Constellation fleet.

LANCASTER: THE MASTER BOMBER

LANCASTER and BOMBER COMMAND

The arrival into squadron service of the first Lancasters in late 1941/early 1942 made a substantial contribution to the improved effectiveness of RAF Bomber Command. Here was a heavy bomber of genuine class, one which was superior to the existing four engined 'heavies' – the Shorts Stirling and Handley Page Halifax – and one regarded by many as simply the best bomber of World War II, the Boeing B-29 not withstanding.

Bomber Command's strength at the outbreak of war in September 1939 comprised 53 squadrons including 20 non operational units which were in effect reserve squadrons. The aircraft available to the then Commander-in-Chief, Air Chief Marshal Sir Edgar Ludlow-Hewitt, ranged from the next to useless Fairey Battle single engined day bomber, through to various light and medium twins: the Bristol Blenheim, Armstrong Whitworth Whitley, Handley Page Hampden and Vickers Wellington. Of these, the Wellington was the best, and carried the brunt of Bomber Command's activities for the first two years of the war until the four engined bombers became available in numbers.

By the standards of the day, the Wellington offered useful performance and was able to carry a 4,500lb (2,040kg) bomb load over a range of 1,500 miles (2,410km). This payload/range performance wasn't far short of the four engined Boeing B-17E Fortress, which could travel just under 2,000 miles (3,220km) with a 4,000lb (1,814kg) load.

Hard Lessons

Like their colleagues in the USA, the British believed that the bomber alone could turn a war into victory by destroying the enemy's industrial base and its citizens' morale. There was also a strong belief that the "bomber will always get through", that an airborne armada would be able to fight its way to any target and fight its way back home again, defending itself all the way.

Although the British and American bombing campaigns against Germany in World War II proved to be an essential part of the ultimate victory, they were not the only element. Industrial centres which sustained serious damage tended to be put only temporarily out of action and the morale of the recipients of the bombing rarely wavered. The same applied to Britain when the *Luftwaffe* was attacking it during the summer of 1940 and afterwards.

The early months of the war saw Bomber Command restrained from attacking anything other than purely military targets in fear of a savage response from the *Luftwaffe*. There was a reluctance even to attack the industrial centres of the Rühr Valley – the vital hub of the German war machine – because of this and an almost benevolent attitude by many, including the Air Minister, Sir Kingsley Wood. When the idea of bombing the Rühr was suggested to him, Sir Kingsley rejected it because the factories were private property!

Early raids were therefore conducted mainly against warships and in daylight. Losses were alarmingly high and results poor, forcing a

"Fill 'er up, 2,154 gallons please". A 463 Squadron Lancaster is refuelled prior to another mission from Waddington. (via Neil Mackenzie)

switch to minelaying ('Gardening') and leaflet dropping ('Nickelling') operations. One of these to Berlin in late 1939 "failed to shame Hitler and cronies into surrendering," it was dryly noted, despite the leaflets containing information about how the Nazi leaders had removed large amounts of money from Germany to safe overseas locations!

More serious was the fact that the early combat operations that revealed no matter how many gun turrets a bomber might have, it could not fight its way to and from a target in daylight without fighter escort. Ludlow-Hewitt blamed the problems on "poor leadership and consequent poor formation flying" and stated there was "every reason to believe that a very close formation of six Wellington aircraft will emerge from a long and heavy attack by enemy fighters with few, if any, casualties to its own aircraft". At least the powers-that-be recognised the need for self sealing fuel tanks and increased armour protection for the bomber crews and did something about them.

Dropping leaflets instead of bombs on Germany did little for the crews' morale, did no damage to the enemy and cost valuable aircraft. One important change to Bomber Command's general philosophies did emerge from those first few months of the war, and that was a shift in emphasis from day to night operations. The second major change which would influence the Lancaster's and other RAF bombers' operational careers was the switch to area bombing, but that would not occur until 1942.

Operations against targets on dry land finally got underway in March 1940 and against German territory two months later. The first of them

Air Chief Marshal Sir Arthur Harris, AOC-in-C RAF Bomber Command February 1942 to September 1945.

were against oil and rail targets in the Rühr but these and subsequent raids against industrial targets soon revealed very poor results as bombing accuracy was almost non existent. By now, Air Marshal Sir Charles Portal was in charge of Bomber Command and he gradually introduced changes in philosophy which would see RAF bomber operations evolve towards area bombing.

From mid 1940 the bombers began to concentrate on defined priority targets: initially aircraft assembly plants, aircraft depots, oil refineries and with a view to "complete destruction rather than harassing effect". Naval targets – particularly U-boat yards and bases – became a very high priority from 1941 as the Battle of the Atlantic became critically important.

A reluctance to attack targets close to civil populations remained, particularly in view of the expected inaccuracy of the bombing, remembering these were the days before radar and other electronic assistance in that area and that once a target was covered by cloud it was in effect invisible. Even in bright moonlight Portal was of the opinion that only three out of ten targets would be found with any certainty by the average crew and that the very high percentage of bombers which inevitably miss the target will hit nothing else (ie civilians) and do no damage.

Harris and the Heavies

Portal was promoted to Chief of Air Staff in October 1940 and replaced by Air Marshal Sir Richard Peirse as C-in-C Bomber Command. Peirse held the position until January 1942, Air Vice-Marshal J E A Baldwin acting in the role until the following month when Air Chief Marshall Sir Arthur Harris took over. He remained Bomber Command's boss until the end of the war.

'Bomber' Harris' arrival coincided with the formation of the earliest Lancaster squadrons and a situation in which Bomber Command's value was being questioned in many quarters. 1941 had not been a great year for the Command. Although it had seen the introduction to service of the RAF's trio of four engined heavy bombers (the inadequate due to its small wing Stirling early in the year, the better Halifax in March and the Lancaster in December), there were serious problems due to a lack of

Lancaster I ME844/LS-W of 15 Squadron in 1944.

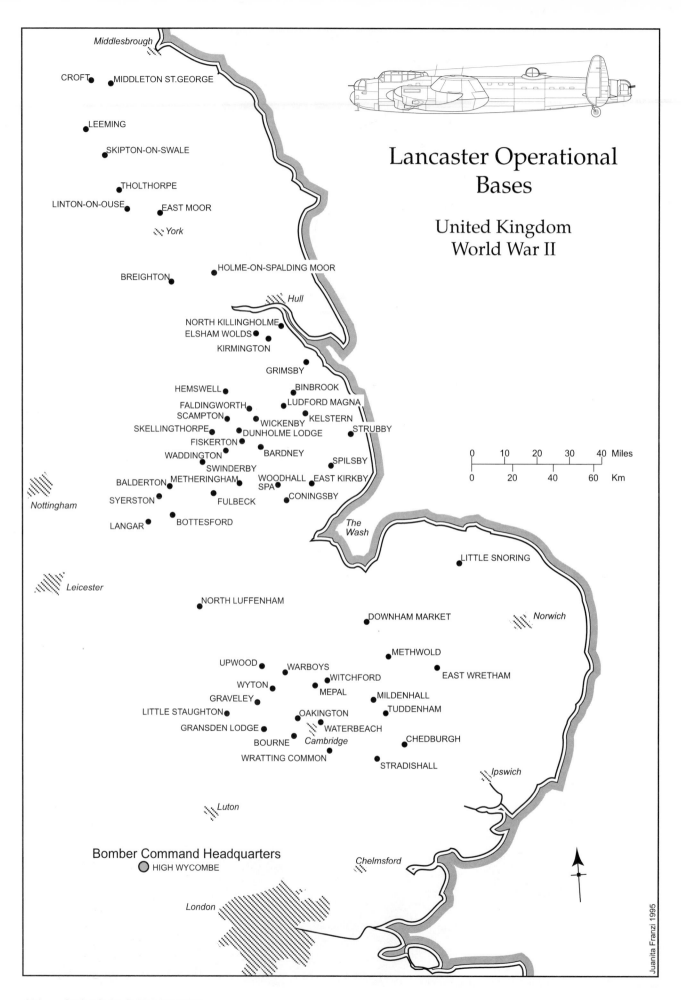

Lancaster Operational Bases

United Kingdom World War II

Middlesbrough

CROFT
MIDDLETON ST.GEORGE

LEEMING

SKIPTON-ON-SWALE

THOLTHORPE

LINTON-ON-OUSE
EAST MOOR

York

BREIGHTON
HOLME-ON-SPALDING MOOR

Hull

NORTH KILLINGHOLME
ELSHAM WOLDS
KIRMINGTON

GRIMSBY

HEMSWELL
BINBROOK
FALDINGWORTH
LUDFORD MAGNA
SCAMPTON
KELSTERN
SKELLINGTHORPE
WICKENBY
DUNHOLME LODGE
STRUBBY
FISKERTON
BARDNEY
WADDINGTON
SPILSBY
SWINDERBY
BALDERTON
METHERINGHAM
WOODHALL
EAST KIRKBY
SYERSTON
SPA
FULBECK
CONINGSBY

LANGAR
BOTTESFORD

Nottingham

Leicester

The Wash

LITTLE SNORING

NORTH LUFFENHAM

DOWNHAM MARKET
Norwich

METHWOLD

UPWOOD
WARBOYS
WITCHFORD
EAST WRETHAM
WYTON
MEPAL
GRAVELEY
MILDENHALL
LITTLE STAUGHTON
OAKINGTON
TUDDENHAM
GRANSDEN LODGE
WATERBEACH
CHEDBURGH
BOURNE
Cambridge
WRATTING COMMON
STRADISHALL

Ipswich

Bomber Command Headquarters
⊙ HIGH WYCOMBE

Chelmsford

Luton

London

0 10 20 30 40 Miles
0 20 40 60 Km

Juanita Franzi 1995

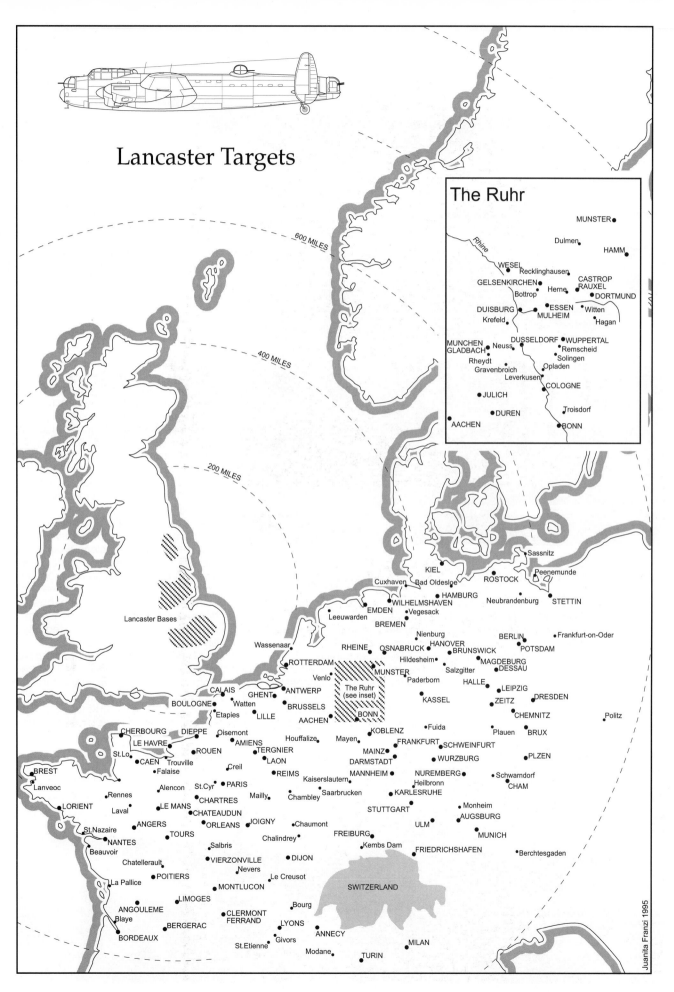

Lancaster Targets

The Ruhr

MUNSTER
Dulmen
HAMM
WESEL
Recklinghausen
CASTROP
RAUXEL
GELSENKIRCHEN
Herne
Bottrop
DORTMUND
ESSEN
Witten
DUISBURG
MULHEIM
Krefeld
Hagan
MUNCHEN
DUSSELDORF
WUPPERTAL
GLADBACH
Neuss
Remscheid
Rheydt
Solingen
Gravenbroich
Opladen
Leverkusen
COLOGNE
JULICH
Troisdorf
DUREN
AACHEN
BONN

Rhine

600 MILES

400 MILES

200 MILES

Lancaster Bases

Sassnitz
Peenemunde
KIEL
ROSTOCK
Cuxhaven
Bad Oldesloe
Neubrandenburg
STETTIN
HAMBURG
WILHELMSHAVEN
EMDEN
Vegesack
Leeuwarden
BREMEN
Nienburg
BERLIN
Frankfurt-on-Oder
HANOVER
Wassenaar
RHEINE
OSNABRUCK
BRUNSWICK
POTSDAM
Hildesheim
MAGDEBURG
ROTTERDAM
MUNSTER
Salzgitter
DESSAU
Venlo
Paderborn
HALLE
CALAIS
GHENT
ANTWERP
The Ruhr
(see inset)
KASSEL
LEIPZIG
BOULOGNE
Watten
BRUSSELS
ZEITZ
DRESDEN
Etapies
LILLE
AACHEN
BONN
CHEMNITZ
Politz
CHERBOURG
DIEPPE
Oisemont
KOBLENZ
Fuida
BRUX
LE HAVRE
AMIENS
Houffalize
Mayen
FRANKFURT
Plauen
St.Lo
ROUEN
TERGNIER
MAINZ
SCHWEINFURT
PLZEN
BREST
CAEN
Trouville
LAON
DARMSTADT
WURZBURG
Falaise
Creil
REIMS
Kaiserslautern
MANNHEIM
NUREMBERG
Schwarndorf
Lanveoc
Alencon
St.Cyr
PARIS
Saarbrucken
KARLSRUHE
CHAM
Rennes
Mailly
Chambley
Laval
LE MANS
CHARTRES
STUTTGART
Monheim
LORIENT
CHATEAUDUN
ULM
AUGSBURG
St.Nazaire
ANGERS
ORLEANS
JOIGNY
FREIBURG
MUNICH
Chaumont
NANTES
Chalindrey
Kembs Dam
FRIEDRICHSHAFEN
Beauvoir
TOURS
Berchtesgaden
Salbris
Chatellerault
VIERZONVILLE
DIJON
Nevers
SWITZERLAND
La Pallice
POITIERS
Le Creusot
MONTLUCON
Bourg
LIMOGES
ANGOULEME
CLERMONT
FERRAND
Blaye
LYONS
BERGERAC
ANNECY
BORDEAUX
Givors
MILAN
St.Etienne
Modane
TURIN

Juanita Franzi 1995

Lancaster on short finals with its large flaps well displayed.

(left) The office. Although the Lancaster could be fitted with dual controls by extension arms for the control column and rudders, very few were so equipped. The standard Lancaster operational crew had one pilot only.

confidence in the Command thanks to its poor performance.

There was increasing pressure to divert more of Bomber Command's resources to Coastal Command in order to help in the desperately important fight against the U-boats, this against a background of heavy criticism of the bomber force. A report into Bomber Command's performance against targets in France and Germany revealed that in June and July 1941 only one in three bomber crews had been able to place their bombs within five miles (8km) of the aiming point! Against targets in the Rühr Valley the figure was even worse with just one in five achieving the same level of 'accuracy'.

The report also showed that bright moonlight was an indispensable factor in the crews achieving any sort of accuracy at all. The figures improved to 40 per cent of crews placing their bombs within five miles of the target on moonlit nights but this was largely countered by the increasing effectiveness of the German night fighter organisation. Losses began to mount, criticism began to increase.

Various plans were put in place to send ever larger formations of bombers to Germany to flatten it with high explosives. Portal argued for increased bomber production, saying that 4,000 bombers could remove 43 German cities from the map in the space of six months. Prime Minister Winston Churchill (who never believed that bombing alone would win the war but nevertheless appreciated its importance as the only weapon he could use to strike at the heart of Germany) decided to scale operations down over the northern hemisphere winter of 1941-42, regroup, and start again in the spring.

In the meantime there would be a new Commander-in-Chief of Bomber Command (Harris), new and more effective aircraft (Lancasters and Halifaxes) and a new attitude. One of Harris' earliest moves was to order

Waiting its turn. Although this Lancaster still has its wheel covers on, bomb doors open and a maintenance platform is in place under the port inner engine, a fully kitted crew appears ready to climb aboard. (via Neil Mackenzie)

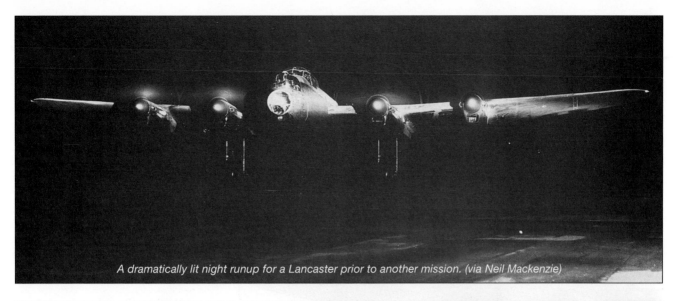

A dramatically lit night runup for a Lancaster prior to another mission. (via Neil Mackenzie)

Lancaster B.III PB509/OJ-C of 149 Squadron at Methwold. The stripes on the fin denote a Gee-H equipped formation leader. (via Neil Mackenzie)

Bombing up with short finned 1,000 pounders, lifted aboard via a winching system. (via Neil Mackenzie)

Taxying out. Lancasters of 467 (RAAF) Squadron at Waddington in 1945. (via Neil Mackenzie)

(left) Another Gee-H equipped formation leader, this time B.I NG299 of 149 Squadron. (via Neil Mackenzie)

the first '1,000 Bomber' raid of the war against Cologne (Operation Millennium) on the night of 30/31 May 1942. The fleet of bombers and their crews were scraped together from every available source; not just from the squadrons but from reserve and training units as well. Most were Wellingtons and other twin engined bombers but a small number of the new Lancasters also participated.

The 'Thousand Bomber Raid' (1,047 aircraft departed of which an estimated 870 actually bombed the target) had great benefits both to Bomber Command's until then questionable future and to Britain's morale. Harris chose the figure '1,000' for the number of bombers he wanted to attack Cologne mainly for its publicity value – it rolled off the

tongue and onto the front pages quite nicely – but the fact that Bomber Command could actually assemble a fleet of that many aircraft for a single raid quietened its critics somewhat. British morale rose but Germany's must have sunk slightly with the realisation that the Fatherland was not immune to large scale attack from the air.

Area Bombing

The Cologne raid was an early indication of a new direction for Bomber Command. Recognising the problems with bombing accuracy, Harris directed that a new policy – area bombing – would come into force from March 1942, naming major German cities as the primary targets. This policy specifically directed that

aiming points would be built up areas and the intention was to damage the enemy civilian population's morale as much as its industrial and military facilities. Studies into the subject seemed to confirm that the destruction of dwellings had a greater effect on the lowering of civilian morale than even the deaths of friends and family by bombing.

To make this type of bombing as effective as possible, a new range of weapons had been developed. Studies had shown that before 1942, British bombs packed less punch than their German equivalents, so work was put in hand to improve the situation. The result was a series of bombs with a very high explosives contents, just the thing for the demolition work which formed the basis of the area bombing theory.

The amount of metal in each bomb was reduced in order to increase the amount of explosives which could be carried for a given weight, the result being light case, high capacity bombs such as the 4,000lb (1,814kg) 'Cookie', which externally looked just

Armstrong Whitworth built Lancaster I LM233 of 467 Squadron taxies out. (via Neil Mackenzie)

like an enlarged 44 gallon drum. Smaller, more conventionally shaped bombs also got a higher explosives content than before (about 40 per cent for the 500-pounder) but the entirely new weapons such as 'Cookie' were a remarkable 80 per cent explosives.

By the end of the European war in May 1945, about 75 per cent of Bomber Command's bombs had fallen on what could be termed 'area' targets. Despite this general philosophy, the history of the Command's operations during World War II is full of examples of raids conducted against specific military and industrial targets, raids which required an extraordinarily high degree of precision and innovation, often using special weapons and training methods. Most involved the Lancaster and include such celebrated efforts as the Dams raid in May 1943 with the special 'Upkeep' bouncing bomb.

Pathfinder Force

The need for more accurate bombing was an urgent requirement by the time Arthur Harris took charge of Bomber Command. One of his more important decisions was to establish a special force of 'target find-

Almost as important as the production lines building new Lancasters were the workshops repairing damaged ones and returning them to service. The damaged aircraft were broken down into major components, the repairs carried out and then the parts bolted together again. These nose sections are undergoing refurbishment at Langar, one of Avro's own repair shops, in 1943.

ers' to lead the main bomber streams and accurately mark the targets for those which followed.

Although Harris supported the general concept, he initially opposed the idea of establishing what would in effect be an elite corps within Bomber Command, arguing that a better method would be to employ one squadron from within each group to do the job. Strong argument from various quarters (including from the Chief of Air Staff and various of the War Cabinet) resulted in Harris being overruled and the decision to establish the Pathfinder Force (initially as

Famous last words. Herman Goering's boast probably still haunts him, even in the afterlife! The Lancaster is the famous R5868/PO-S 'S for Sugar' of 467 (RAAF Squadron), a veteran of 137 missions and now preserved in the RAF Museum, Hendon. When this photograph was taken in late 1944 'Sugar' had logged 118 ops. (via Neil Mackenzie)

An atmospheric shot of R5868 'S-Sugar' on the ground. (via Neil Mackenzie)

part of No 3 Group and from early 1943 its own Group, No 8) was made in June 1942. To his credit, Harris accepted the decision which went against his own wishes and fully supported this new *corps d'elite*.

Grp Capt (later AVM) Donald Bennett, an Australian well known for his pre war long distance flights, was selected to command the Pathfinder Force which initially comprised Nos 7 (Stirlings), 35 (Halifaxes), 83 (Lancasters) and 156 (Wellingtons) Squadrons. As the war progressed, the PFF evolved into a two aircraft organisation using Lancasters and Mosquitos, the latter's small size, ability to carry a 4,000lb 'Cookie' bomb and high speed making it ideal for the role. By the end of the European war the PFF had 16 squadrons on strength, 10 equipped with Mosquitos and six

(Nos 7, 35, 156, 405 RCAF, 582 and 635) with Lancasters.

The PFF's early role was principally finding and illuminating targets by using specially developed bombs containing various incendiary devices which burnt in a particular colour and/or pattern and were easily recognisable by the following bombers. Three basic marking techniques were employed: visual groundmarking (codenamed 'Newhaven'), blind groundmarking ('Parramatta' after Bennett's home suburb in western Sydney) and skymarking ('Wanganui', after one of the squadron commander's New Zealand hometown).

Pathfinding required the highest possible skills if the targets were to be marked accurately. It was all done visually in 1942 but the arrival of H2S ground mapping radar and the use of

Oboe equipment as a blind bombing device further improved the already extraordinarily high standards being displayed by the PFF crews.

It was hazardous work, too. PFF aircraft flew 50,490 individual sorties against 3,440 targets at a cost of 3,727 men killed on operations, the equivalent of 20 Lancaster squadrons.

One other technique developed to improve overall bombing accuracy by a large force was that of the Master Bomber. Intended to act as a kind of 'master of ceremonies', the Master Bomber remained over the target throughout the raid, giving instructions to the main force as to where the bombs should be placed so as to achieve maximum concentration. This in combination with accurate marking by the PFF gave the bombers a much improved chance of hitting the target.

The first (unplanned) use of what would evolve into the Master Bomber technique was by 617 Squadron on the Dams raid in May 1943 while its first use by a major force was against Turin the following August.

Over the last year of the war the standard of bombing accuracy achieved by Bomber Command was at a level inconceivable two or three years earlier. That in combination with diminishing resistance from enemy defences from late 1944 and ever increasing numbers of bombers – both British and American – ensured that Germany would suffer great levels of destruction.

Continued Expansion

Bomber Command expanded rapidly from 1943. By March of that year it had 65 Squadrons of which 37 were four engined 'heavies'. Of these, 20 were equipped with Lancasters. By the end of hostilities in Europe, 61 Squadrons had received Lancasters of which one had been

A Lancaster of 101 Squadron drops a load of incendiary bombs over Duisberg in October 1944. A 4,000lb 'Cookie' would have also been part of the load. Note the aircraft's ABC jamming equipment ('Cigar') aerials on the upper fuselage and the fact that this a daylight mission, something which became more commonplace after D-Day.

formed in late 1941, 15 in 1942, 20 in 1943, 19 in 1944 and six in 1945.

The Commonwealth nations made a significant contribution to Bomber Command's strength through the Empire Air Training Scheme, the plan under which tens of thousands of airmen were trained for RAF service mainly in Canada, Australia, Rhodesia and New Zealand. After training, airmen were posted to the RAF, often in newly formed squadrons which reflected their nationality and were designated as such. Among the Lancaster

Lancaster B.III LM446/PG-H of 619 Squadron. (via Neil Mackenzie)

RAF LANCASTER SQUADRONS 1941-1954

Notes: The table lists RAF Lancaster squadrons (including postwar) in numerical order including those Royal Australian Air Force (RAAF), Royal Canadian Air Force (RCAF) and Royal New Zealand Air Force (RNZAF) units which flew under the control of RAF Bomber Command. The 'dates' column lists the period of time a particular squadron had Lancasters on strength. The postwar squadrons have the type of Lancaster operated noted if it's a mark other than a standard one.

Sqdn	Codes	Dates	Bases/Remarks	Sqdn	Codes	Dates	Bases/Remarks
7	MG	05/43-08/49	Oakington, Mepal, Upwood	189	CA	10/44-11/45	Bardney, Fulbeck, Metheringham
9	WS	08/42-04/46	Waddington, Bardney, India	195	AC/JE	10/44-08/45	Witchford, Wratting Common
12	PH	11/42-08/46	Wickenby, Binbrook	203		05/47-03/53	GR.III; St Eval, Topcliffe
15	LS	12/43-03/47	Mildenhall, Wyton	207	EM	04/42-09/49	Bottesford, Langar, Spilsby,
18		09/46 only	GR.III, Palestine				Methwold, Tuddenham, Stradishall,
35	TL	03/44-10/49	Graveley, Stradishall, Mildenhall				Mildenhall
37		04/46-08/53	B.III/B.VII/MR.III; Egypt, Palestine,	210	OZ	06/46-02/53	ASR.III/MR.III; St Eval, Topcliffe
			Malta	214	QN	11/45-02/50	Egypt, Upwood
38	RL	07/46-12/54	ASR.III/MR.III; Malta, Palestine	218	HA	08/44-08/45	Methwold, Chedburgh
40	BL	01/46-04/47	B.VII, Egypt	224	XB	10/46-11/47	GR.III, St Eval
44	KM	12/41-05/47	first Lanc squadron; Waddington,	227	9J	10/44-09/45	Bardney, Balderton, Srubby, Graveley
			Dunholme Lodge, Spilsby, Mepal,	279		09/45-03/46	ASR.III, Beccles
			Mildenhall, Marham	300	BH	04/44-10/46	Polish; Faldingworth
49	EA	07/42-10/49	Scampton, Fiskerton, Fulbeck,	405	LQ	08/43-06/45	RCAF; Gransden Lodge, Linton-on-
			Syerston, Mepal, Upwood				Ouse; returned to Canada, kept
50	VN	05/42-11/46	Skellingthorpe, Swinderby, Sturgate,				Lancs until 1955
			Waddington	408		08/43-08/44	RCAF, Mk.II only, Linton-on-Ouse
57	DX	09/42-05/46	Scampton, East Kirkby, Elsham Wolds	419		03/44-06/45	RCAF; Middleton St George
61	QR	05/42-05/46	Syerston, Skellingthorpe, Coningsby,	420		04/45-06/45	RCAF; Tholthorpe
			Sturgate, Waddington	424	QB	01/45-10/45	RCAF; Skipton-on-Swale
70		04/46-04/47	Egypt	425		05/45-06/45	RCAF; Tholthorpe
75	AA	03/44-10/45	RNZAF; Mepal, Spilsby	426	OW	06/43-05/44	RCAF; Mk.II only; Linton-on-Ouse
82		10/46-12/53	PR.I; Benson, Wyton, Kenya, Gold	427	ZL	03/45-05/46	RCAF; Leeming
			Coast	428		06/44-06/45	RCAF; Middleton St George
83	OL	04/42-07/46	Coningsby, Wyton, Hemswell	429	AL	03/45-05/46	RCAF; Leeming
90	WP	05/44-12/47	Tuddenham, Wyton	431		10/44-06/45	RCAF; Croft
97	OF	01/42-07/46	Coningsby, Woodhall Spa, Bourn	432		10/43-02/44	RCAF; Mk.II only; East Moor
100	HW	01/43-05/46	Grimsby, Elsham Woods, Scampton	433	BM	01/45-10/45	RCAF; Skipton-on-Swale
101	SR	10/42-08/46	Holme-on-Spalding Moor, Ludford	434	IP	12/44-06/45	RCAF; Croft
			Magna, Binbrook	460	AR	10/42-08/46	RAAF; Breighton, Binbrook, East Kirkby
103	PM	11/42-11/45	Elsham Wolds	463	PO/JO	11/43-09/45	RAAF; Waddington, Skellingthorpe
104	EP	11/45-04/47	B.VII, Egypt	467	PO	11/42-09/45	RAAF; Scampton, Bottesford,
106	ZN	04/42-02/46	Coningsby, Syerston, Metheringham				Waddington, Metheringham
109		08/42 only	brief trials only at Wyton	514	JI/A2	09/43-08/45	Foulsham, Waterbeach
115	KO/IL	03/43-09/49	East Wretham, Little Snoring,	541		05/46-09/46	Benson, planned survey work in
			Witchford, Graveley, Stradishall,				Africa
			Mildenhall	550	BQ	10/43-10/45	Grimsby, North Killingholme
120	BS	11/46-04/51	GR.III/ASR.III; Leuchars, Kinloss	576	UL	11/43-09/45	Elsham Wolds, Fiskerton
148	AU	11/46-01/50	Upwood	582	60	04/44-09/45	Little Staughton
149	OJ	08/44-11/49	Methwold, Tuddenham, Stradishall,	617	AJ/KC	03/43-09/46	Scampton, Coningsby, Woodhall Spa,
			Mildenhall				Binbrook, India; also YZ code
150	IQ	11/44-11/45	Fiskerton, Hemswell	619	PG	04/43-07/45	Woodhall Spa, Coningsby, Dunholme
153	P4	10/44-09/45	Kirmington, Scampton				Lodge, Strubby, Skellingthorpe
156	GT	12/42-09/45	Warboys, Upwood, Wyton	621		04/46-09/46	ASR.III, Palestine
160	BS	08/46-10/46	Leuchars, renamed 120 Sqdn	622	GI	12/43-08/45	Mildenhall
166	AS	09/43-11/45	Kirmington	625	CF	10/43-10/45	Kelstern, Scampton
170	TC	10/44-11/45	Kelstern, Dunholme Lodge, Hemswell	626	UM	11/43-10/45	Wickenby
178		11/45-04/47	Egypt	630	LE	11/43-07/45	East Kirkby
179		02/46-09/46	ASR.III, St Eval	635	F2	03/44-09/45	Downham Market
186	AB	10/44-07/45	Tuddenham, Stradishall	683		11/50-11/53	PR.I; Egypt, Iraq

A selection of British World War II bombs, ranging from the 40lb (18kg) general purpose weapon on the airman's shoulder to the 22,000lb (9,980kg) Grand Slam, the most powerful conventional bomb of the war.

Cookies for Adolf, with compliments from the Australians serving in Bomber Command. (via Neil Mackenzie)

squadrons there were 14 Royal Canadian Air Force (RCAF – mostly flying Canadian built Lancaster Xs) and three Royal Australian Air Force (RAAF) units numbered in the '400' series. Many thousands of airmen from these and other countries also flew with the 'normal' RAF squadrons in all Commands.

Early Operations

The Lancaster force had grown quickly since December 1941 when the first operational squadron equipped with the new bomber – No 44 (Rhodesian) Squadron at Waddington – had taken it into service. To 44 Squadron also went the honour of conducting the first Lancaster operations, the initial mission occurring on 3 March 1942 when four aircraft undertook a minelaying task off the German coast.

The Lancaster's first bombing mission was recorded on the night of 10/11 March when two of 44 Squadron's Lancasters participated in a raid on Essen. Both returned safely.

The second squadron to receive Lancasters was No 97 at Coningsby in January 1942. It also began operations with a mining sortie (a night mission by two aircraft to the Baltic on 20 March), while its maiden bombing raid was carried out on 25/26 March when seven Lancasters from it and 44 Squadron joined a raid on Essen. Once again, all returned safely, although 44 Squadron had recorded its (and the Lancaster's) first operational loss the night before when one of the two aircraft sent out on a mining sortie failed to return.

A major early raid in which Lancasters participated was that on the MAN diesel engine factory at Ausberg on 17 April, by the standards of the day a long range mission involving a 1,000 miles (1,610km) round trip and complicated navigation. The target was important because it was a major manufacturing site for tank and U-boat engines, the latter causing Britain substantial problems in the Atlantic at the time. The fact that the raid was conducted in daylight was significant to its outcome.

Twelve Lancasters took part, six each from 44 and 97 Squadrons. The trip to the target was mostly flown at very low level but the formation encountered heavy opposition from German fighters, which shot down four Lancasters before the target was reached.

Three more were dispatched in the target area and of the five Lancasters which made it home, one was written off due to the extreme damage it had suffered. Despite this, the raid was in one sense successful because the target had been badly damaged, but

Vickers-Armstrong built Lancaster I PA238/SR-Z of 101 Squadron during its takeoff run.

Lancaster I W4119/VN-Q of 50 Squadron early in 1944, shortly before it caught fire in the air and crashed near East Kirkby, Lincolnshire. Note the open crew door. (via Neil Mackenzie)

The V-2 rocket centre at Peenemünde before (top) and after the attentions of Bomber Command.

The winter of 1944-45 was a cold one but come the spring, these snow covered Lancasters would see the end of hostilities. (via Neil Mackenzie)

the overall feeling was of misgiving because of the very high loss rate. This was largely caused by the fact that the mission was flown in daylight, a situation which contradicted the lessons Bomber Command had previously learned.

In the PR sense, the Ausberg raid was successful in that its daring gave the British people a boost when it was badly needed.

Bigger and Further

1943 was a big year in the Command's history with the bomber offensive against Germany building up to the massive operation it became. The USAAF's 8th Air Force also began to get into its stride during 1943 (albeit with very heavy losses throughout the year), it's daylight operations complementing the RAF's night raids. Strangely, there was little co-opera-

tion between the British and American forces, their campaigns usually running entirely separately from each other.

Harris launched the Battle of Rühr – the campaign against Germany's industrial hub – in March 1943, the first phase of this battle ending two months later after some intensive operations against centres such as Essen, Duisberg, Dusseldorf, Dortmund and Bochum. Other significant 1943 operations included the dams raid in May; Operation Gemorrah – the devastation of Hamburg – started in July; the 600 bomber raid on the V-2 rocket research site at Peenemunde in August (which involved an early use of the Master Bomber technique; the Dortmund-Ems Canal raid by 617 Squadron in September (including the first use of the 12,000lb/5,448kg 'Tallboy' deep penetration bomb;

campaigns against major cities such as Munich, Stuttgart, Mannheim, Frankfurt, Leipzig, Brunswick and Nuremburg; and the start of the Battle of Berlin in November.

The Big City

The first major assault on Germany's capital took place on the night of 18 November 1943 and involved 416 Lancasters from 26 squadrons of which 405 reached Berlin and dropped 1,590 tonnes of bombs on the city for the loss of 11 aircraft. The raid was not entirely successful in the sense that the bombs were scattered over a much larger area than had been intended due to inaccurate marking.

The Battle of Berlin continued until March 1944 and cost Bomber Command 492 aircraft missing and nearly 1,000 damaged. It witnessed the use

Moving the engine stands up to a Lancaster prior to engine maintenance in the snow and slush of late 1944. (via Neil Mackenzie)

of numerous electronic and mechanical devices by both sides as the need for greater navigation and bombing accuracy grew as did the skill of the German night fighters. Apart from radar, the Lancasters used most of the 'black box' equipment described in the previous chapter and the simple 'Window' aluminium strips designed to swamp German radar.

Arthur Harris' decision to mount a major campaign against Berlin resulted from his continuing belief that constant attacks on the civilian population would at least reduce the Germans' will to a point that would make the planned invasion of Europe by the Allies less costly and quicker. In November 1943 he told Winston Churchill: "We can wreck Berlin from end to end if the US Army Air Force will come in on it. It will cost us between 400 and 500 aircraft. It will cost Germany the war".

As noted above, the actual cost was 492 aircraft following 16 major raids on Berlin comprising 9,111 individual sorties. The loss rate was therefore 5.4 per cent, higher than average, and the city presented a difficult target as it was a long way away, was very much larger than other cities which had been attacked and was heavily defended. The most elaborate defensive measure was provided by the town of Nauen, 15 miles (24km) west of Berlin which was used as a 'dummy' Berlin to confuse the bombers and reduce their accuracy. Nauen was lit up with searchlights, fires, fake bomb flashes and false target indicators, all de-

A spot of bother for a 582 Squadron Lancaster. Note how the liferaft has popped out of its compartment in the starboard wing root. (via Neil Mackenzie)

signed to make the bomber crews think it was really Berlin!

Another problem facing those who thought that Berlin could be destroyed by bombing was the fact that the raids were conducted with insufficient intensity. Berlin was of such a size that it needed to be attacked in large numbers on an almost a round-the-clock basis if it was to be severely damaged. Sixteen RAF raids over a four month period on a large area was never going to destroy the city, even with the help of the Americans during the day.

Fighter Opposition

The cat and mouse game between the RAF bombers and *Luftwaffe* night fighters provoked one of the areas of greatest technology growth in the war as measure and countermeasure were played off against each other. In the 12 month period between April 1943 and March 1944 Bomber Com-

mand lost 2,700 aircraft and 19,000 aircrew, most of them to German night fighters. The November 1943 to March 1944 period alone saw more than 1,000 RAF bombers fall to the night fighters.

The German night fighters had effective *Lichenstein* radar in their Messerschmitt Bf 110s and Junkers Ju 88s plus the *Schräge Musik* system designed to exploit the lack of ventral guns on the British heavy bombers. *Schräge Musik* literally means 'oblique music' – jazz, in other words – and as applied to the German night fighters was a pair of 20mm or 30mm cannon installed in the upper fuselage of the fighter with the guns pointing upwards and slightly forwards.

Use of this effective weapon involved the German pilot manoeuvring his fighter to a position underneath his target bomber, approaching from an angle which prevented him from

Lancasters of 467 Squadron at Waddington. (via Neil Mackenzie)

wandering into the area protected by the bomber's Monica tail warning radar or the rear gunner. Once installed in that position, the fighter was impossible to see from the bomber while it presented a nice silhouette to the fighter pilot against the sky above.

Many bombers were lost to this method, most of them not knowing what had hit them and it took several months of *Schräge Musik* attacks before the British were able to work out what was happening. The removal of the Lancaster's ventral gun position early in its career must have been a cause for regret at this time, although some units (particularly the Canadian squadrons) kept theirs.

Towards Victory

Bomber Command underwent a reorganisation in April 1944 which saw it come under the control of the Supreme Allied Commander, General Eisenhower. The style of operations changed at the same time with transportation systems and facilities in France regarded as a priority preliminary to the Allied invasion at Normandy the following June. The seven week period leading up to the invasion saw Bomber Command drop some 42,000 tonnes of bombs on the French transportation system and effectively reduce it to a pile of scrap.

The D-Day landings on 6 June were supported by a large bomber assault the day and night before with ten German coastal batteries in the Normandy area put out of action after receiving the attentions of more than 1,100 heavy bombers which between them dropped nearly 5,300 tonnes of bombs. On D-Day and over the days which followed, Bomber Command concentrated on German supply routes and installations in the path of the invading armies.

The strategic bombing campaign was resumed after that, with the Lancasters and other British heavy bombers operating more often in daylight as the Allies gradually gained air supremacy. The last year of the war witnessed greater co-operation between the RAF and USAAF bomber fleets as German oil and aircraft manufacturing facilities came in for special attention. Germany's oil industry was systematically destroyed over the final six months of the war, and this was probably the Lancaster's and other bombers' greatest single contribution to the circumstances which resulted in Germany's surrender. Starved of fuel, the German war machine simply could not operate.

The last bombing missions of the war undertaken by Lancasters occurred on 25 April 1945 when Hitler's retreat at Berchtesgaden was attacked during daylight hours with an escort of USAAF P-51 Mustangs. This was followed by a night mission against U-boat oil storage tanks in Oslo Fjord on the same date.

By the end of the war against Germany Bomber Command had expanded to nearly 100 operational units equipped with more than 2,200 aircraft. Of these squadrons, 57 were equipped with Lancasters out of a total of 61 which had flown the Avro bomber since late 1941. Bomber Command flew 364,500 individual sorties during the conflict, dropped about 955,000 tonnes of bombs and lost 8,325 aircraft and 47,268 aircrew in action. To these figures must be added some 16,000 bombers damaged, 4,200 men wounded on ops and 8,300 killed on non operational duties. A heavy toll.

Postscript

With the end of hostilities in Europe, Bomber Command's Lan-

May 1945: the war in Europe is over. (via Neil Mackenzie)

The first of 54 Lancasters for France's Aeronavale, WU-01, the former RAF B.VII NX613. It was delivered in late 1951.

casters and other types were immediately put to more peaceful uses including the repatriation of Prisoners of War (Operations Exodus and Dodge) and the dropping of food to Dutch communities which were suffering extreme shortages (Operation Manna). The Lancasters used in this operation had five panniers installed in their bomb bays, each holding 70 packs containing up to 25lb (11kg) of various tinned foods. The Lancasters used in this operation flew more than 3,100 sorties over The Netherlands, dropping some 6,600 tonnes of food to the starving Dutch population.

The RAF formed another 11 Lancaster squadrons postwar, equipped mainly with Photographic, General and Maritime Reconnaissance conversions of the bomber (as described in the previous chapter), while other postwar operators included Canada (also previously discussed), Argentina (15 refurbished aircraft supplied from 1948), Egypt (nine delivered in 1950) and France.

France's *Aeronavale* acquired an initial quantity of 54 low time Mks I and VII from Britain in 1951 for maritime reconnaissance and search and rescue duties mainly around its various territories in North Africa and the Pacific. The French Lancasters had their dorsal turrets removed, 400imp gal (1,818 l) fuel tanks could be fitted in the bomb bay, late standard ASV radar was installed under the rear fuselage and an airborne lifeboat could be carried.

The aircraft were serialled WU-01 to WU-54, the prefix standing for 'Western Union', not the famous American telegram company but the name applied to a late 1940s concept for collective military aid and economic co-operation by Western European nations.

A further five Lancasters were added to the *Aeronavale's* inventory in 1954 and although most had been replaced by Lockheed Neptunes by the beginning of the 1960s, a few soldiered on from the base at Noumea in

the Pacific until the last three were finally retired in 1964.

AFTER ME, THE DELUGE

Of all the RAF Lancaster squadrons, the best known is undoubtedly No 617 which became famous for its raid against German dams in the Rühr Valley on the night of 16/17 May 1943 – Operation Chastise. This spectacular example of the art of highly precise bombing involved considerable innovation. Foremost of this was use of the Barnes Wallis designed 'Upkeep' weapon, a cylindrical 'bouncing bomb' weighing 9,250lb (4,195kg) which skipped across the surface of the dam water, struck the dam wall and then sunk to a pre determined depth where it exploded against the wall.

The theory behind this was that if one or more of these dams which provided water for the Rühr industries could be breached, chaos would result and production would be severely disrupted.

The Battle of Britain Memorial Flight's Lancaster I PA474, here carrying the markings of the first Lancaster operational squadron, No 44. (Philip J Birtles)

Twenty-three Lancaster B.IIIs were modified to carry the Upkeep bomb (or mine, to be absolutely correct), these and the other equipment needed in order that the backwards spinning weapon could be dropped at precisely the correct height above the water (60 feet/18m) are discussed in the previous chapter. The precision required also extended to the distance away from the dams the bombs were released. This was 400-450 yards and was measured by a simple triangulation sight which made use of the distance between the towers at each end of the dam.

Some of the 'dambusters' (left to right): Flt Lt R D Trevor-Roper, Flt Lt D J H Maltby, Wng Cdr G P Gibson, Flg Off E C Johnson, Flt Lt H B 'Mickey' Martin, Flg Off H S Hobday.

The weapon underwent extensive trials before it was used on its one and only mission, many of these proving unsuccessful before the final design incorporating the necessary strength had been established.

617 Squadron was formed at Scampton in Lincolnshire in late March 1943. Commanded by Wing Commander Guy Gibson, it was formed especially for the dams raid but afterwards undertook other special missions requiring a high degree of accuracy and using new weapons. Its motto was (and is) *Apres mois, le deluge* – 'after me, the deluge'. Rarely has a squadron motto been more appropriate.

617 Squadron's crews were carefully selected for the task. They were regarded as being among 'the best of the best', although few of them knew the purpose of their mission until very late in the day.

The raiding force of 19 Lancasters was divided into three groups. The main attack force – led by Gibson – consisted of nine aircraft in three groups of three. Their primary target was the Möhne Dam, but if that dam was breached before all nine aircraft had attacked the remainder would go on to the Eder.

The second group comprised five Lancasters, its target was the Sorpe dam. The third group was a mobile reserve of five aircraft intended to follow the main forces two hours behind and would if necessary attack what were called 'last resort' target dams – the Ennepe, Lister and Diemal or the primary targets if they had not been breached.

The flight across Europe was conducted at low level on moonlit night and was hazardous with precise navigation close to enemy airfields, towns and installations necessary. The approaches to the dams were also difficult, not only because of the various geographical barriers, but also due to the worse than expected anti aircraft fire. Losses were inevitable from this and other causes, the final tally being no fewer than eight of the dams raid Lancasters failing to return with the loss of 53 crew members and three others captured by the Germans.

Only one Lancaster from the second wave made it to the target area

due to shootdowns, aborts and misadventure. One aircraft had to abort because it was flying so low it had hit the water, ripping the Upkeep weapon from its mounts. The Lancaster made it back to base but on arrival found it had no hydraulics and therefore no brakes.

The first wave finally breached the Möhne dam after an agonising five Upkeeps had successfully been launched at it. There was a fair degree of frustration at this because some crews thought that only one or two would be required. For a while it seemed that despite all the successful hits, the dam was never going to fail! Guy Gibson 'directed traffic' from above, acting in the first – albeit unplanned – use of the Master Bomber technique. Gibson also accompanied the attacking bombers on their runs, attempting to draw fire away from them.

Three Lancasters from the first wave which did not need to drop their weapons at the Möhne diverted to the Eder which was also successfully breached. Gibson remained in attendance, once again directing traffic and drawing enemy fire. Two weapons were dropped on the Sorpe and one on the Ennepe dam without serious effect. Attacking the Sorpe with the Upkeep weapon was really a waste of time anyway as it was of earth rather than concrete construction and the principal behind the bouncing bomb couldn't be used.

The dams raid caused great excitement at the time and afterwards, graphic photographs of millions of gallons of water rushing down the valleys below them and causing vast amounts of destruction fuelling the excitement. It had certainly been a

Lancaster AR-P of 617 Squadron in December 1943, the aircraft of 'Mickey' Martin. (via Neil Mackenzie)

Group Captain Leonard Cheshire VC.

remarkable achievement for all involved and Guy Gibson was awarded the Victoria Cross in recognition of his gallant leadership.

There has been some questioning of the military value of the dams raid in recent years, in line with a general tendency to 'put down' the efforts of Bomber Command by revisionist historians who tend to ignore the circumstances and attitudes which existed *at the time* a certain event took place.

The simple facts are that the dams raid caused widespread flooding, destruction of property and disruption of rail, canal and road transport. Electricity and water supplies were disrupted, valuable crops lost, many German soldiers diverted from other areas to help protect the dams against possible future attacks and tens of thousands of workers were diverted from other projects to repair not only the dams themselves but also the vast amounts of infrastructure that had been swept away by the water.

Tallboys and Grand Slams

617 Squadron subsequently became intimately associated with special operations requiring a high degree of precision and with more remarkable weapons designed by Barnes Wallis, the 12,000lb (5,443kg) Tallboy and 22,000lb (9,980kg) Grand Slam deep penetration bombs. These were designed to reach a supersonic terminal velocity when dropped from high altitude and penetrate deeply into the ground before exploding. The resulting 'earthquake' effect could be used against heavily fortified facilities such as underground U-boat pens.

617's Lancasters reverted to standard configuration after the dams raid and the squadron went through a period of relative inactivity before restarting operations from a new base at Coningsby (until early 1944) and

then Woodhall Spa. Guy Gibson was posted to Bomber Command Headquarters and among his successors (in November 1943) was Wing Commander Leonard Cheshire, possibly the most accomplished of Bomber Command's pilots and a man whose exploits as a Pathfinder would become legendary. Cheshire was also later awarded the Victoria Cross.

617 Squadron (in conjunction with No 9 Squadron) became something of specialists in Tallboy and Grand Slam operations, one of the first efforts being an attack on the Saumer railway tunnel on 8 June 1944, using Tallboys. The line over which the tunnel was built carried German reinforcements to Normandy and it was successfully blocked on that day.

The German battleship *Tirpitz* had

long been a priority target for the RAF but had not been sunk by September 1944. Lancasters had tried to sink the *Tirpitz* before, as early as late April 1942 when the bomber was only starting its service career. On that occasion Lancasters from Nos 44 and 97 Squadrons had dropped 4,000lb (1,816kg) bombs on the ship without success.

On 11 September 1944, 38 Lancasters from Nos 617 and 9 Squadrons deployed to Yagodnik in the Soviet Union in order to mount an attack on the ship which was then anchored in Norway's Alten Fjord. The attack took place on 15 September, 21 Lancasters (15 armed with Tallboys and six with anti shipping bombs) achieving hits but failing to sink the *Tirpitz*. The ship was moved

'Upkeep' fact and fiction: the real dams weapon (top) aboard Guy Gibson's ED932/AJ-G and the mockup used in the 1953 film 'The Dambusters' (bottom). At that time, details of the so-called 'bouncing bomb' were still secret.

(top) "After me, the flood". The remains of the Möhne Dam after 617 Squadron's raid on the night of 16 May 1943. The breach in the wall was 200 feet (61m) wide. (bottom) The Eder Dam after receiving the attentions of 617 Squadron.

to Haak Island near Tromso in October 1944, bringing it just within range of Lancasters based in Britain.

A second attack was therefore planned, again involving 617 and 9 Squadrons. This time the raid originated at Lossiemouth in Scotland and was conducted in late October. The squadrons between them dropped 32 Tallboys at the *Tirpitz*, but none found their mark.

It was a case of third time lucky on 12 November when two of the 28 Tallboys dropped found their mark and the *Tirpitz* rolled over at its mooring.

Nos 617 and 9 Squadrons were responsible for most of the 854 Tallboys dropped by Lancasters during the war and it was a natural progression for them to use the ten-ton Grand Slam when it became available. This enormous weapon required considerable modification of the Lancaster in order to carry it and 32 aircraft were built in early 1945 specifically for the purpose as Lancaster B.I (Specials).

The first live test drop of a Grand Slam from a Lancaster took place on 13 March 1945 and 617 Squadron took the weapon on its first operational mission the next day against the Bielefeld Viaduct. Only one Grand Slam was taken on that raid, the other 14 Lancasters involved were armed with Tallboys.

The fact that Grand Slam's debut occurred so close to the end of the European war meant that its use was limited and only 41 were dropped in total, all by 617 Squadron aircraft and

successfully against railway bridges and U-boat pens, the latter with reinforced concrete roofs up to 32 feet (7m) thick. As was the case with most of the squadron's operations, a high degree of accuracy was required in order to place the bombs precisely.

(above) Deep penetration: The Lancaster was the only bomber capable of carrying Barnes Wallis' Tallboy and Grand Slam 'earthquake' bombs. (via Neil Mackenzie)

A Lancaster B.I (Special), modified to accommodate the Grand Slam bomb. This aircraft (PD127) belongs to No 15 Squadron which converted to the type after the war had ended. (via Neil Mackenzie)

G FOR GEORGE: A LANCASTER'S LOGBOOK

A veteran of no fewer than 90 bombing missions over Europe, Lancaster I W4783 is also one of the few of its type preserved today. Built by Metropolitan Vickers at its Mosley Road, Manchester works, W4783 was delivered to No 460 (RAAF) Squadron in October 1942 and flew all its missions with this Australian squadron.

Coded AR-G, the Lancaster was and is best known as simply 'G for George'.

As can be seen from the table below, G-George took part in raids to all corners of the Third Reich including against the Ruhr Valley, the V-2 rocket test centre at Peenemunde and 16 trips to Berlin. By the time it was retired from operations in April 1944, George was the only example of 460 Squadron's original complement of Lancasters still serving the squadron.

George developed a reputation as a 'lucky' aircraft having survived so many raids. Damage had been inflicted, but none so serious as that sustained on the ground in July 1943 when a 4,000lb 'Cookie' bomb accidentally dropped from another aircraft. Incendiaries began to burn and the 'Cookie' exploded, setting fire to several Lancasters including George. It took six weeks to repair the damage.

George was one of only two Lancasters which were taken on RAAF charge at home. The other was B.III ED930 which arrived in Australia in early June 1943. Given the RAAF serial number A66-1, this aircraft (known as Q-Queenie) toured Aus-

Op No	Date	Target	Captain	Op No	Date	Target	Captain
1	6 Dec 42	Mannheim	Flt Sgt J A Saint-Smith	46	17 Aug 43	Peenemunde	Flt Sgt H Carter
2	8 Dec 42	Denmark (sea mining)	Flt Sgt A McKinnon	47	22 Aug 43	Leverkusen	Flt Sgt H Carter
3	9 Dec 42	Turin	Flt Sgt A McKinnon	48	23 Aug 43	Berlin	Flt Sgt H Carter
4	17 Dec 42	Denmark (sea mining)	Flt Sgt J A Saint-Smith	49	27 Aug 43	Nuremburg	Flt Sgt H Carter
5	16 Jan 43	Berlin	Flt Sgt J A Saint-Smith	50	30 Aug 43	Munchen Gladbach	Flt Sgt H Carter
6	3 Feb 43	Hamburg	Flt Sgt J A Saint-Smith	51	31 Aug 43	Berlin	Flt Sgt H Carter
7	7 Feb 43	Lorient	Flt Sgt J A Saint-Smith	52	3 Sep 43	Berlin	Flt Sgt H Carter
8	11 Feb 43	Wilhelmshaven	Flt Sgt J A Saint-Smith	53	5 Sep 43	Mannheim	Flt Sgt H Carter
9	13 Feb 43	Lorient	Flt Sgt J A Saint-Smith	54	6 Sep 43	Munich	Flt Sgt H Carter
10	14 Feb 43	Milan	Flt Sgt J A Saint-Smith	55	22 Sep 43	Hanover	Flt Sgt H Carter
11	16 Feb 43	Lorient	Flt Sgt J A Saint-Smith	56	23 Sep 43	Mannheim	Flt Sgt H Carter
12	21 Feb 43	Bremen	Flt Sgt J A Saint-Smith	57	27 Sep 43	Hannover	Sqdn Ldr A Nichols
13	26 Feb 43	Cologne	Flt Sgt J A Saint-Smith	58	29 Sep 43	Bochum	Flt Sgt H Carter
14	28 Feb 43	St Nazaire	Flt Sgt J A Saint-Smith	59	2 Oct 43	Munich	Flt Sgt H Carter
15	1 Mar 43	Berlin	Flt Sgt J A Saint-Smith	60	3 Oct 43	Kassel	Flt Sgt H Carter
16	5 Mar 43	Essen	Flt Sgt J Murray	61	4 Oct 43	Ludwigshaven	Flt Sgt H Carter
17	8 Mar 43	Nuremburg	Flt Sgt J Murray	62	7 Oct 43	Stuttgart	Flt Sgt H Carter
18	9 Mar 43	Munich	Flt Sgt J Murray	63	8 Oct 43	Hanover	Flt Sgt H Carter
19	11 Mar 43	Stuttgart	Wng Cdr C Martin	64	18 Oct 43	Hanover	Flt Sgt H Carter
20	12 Mar 43	Essen	Flt Sgt J Murray	65	20 Oct 43	Leipzig	Plt Off N Peters
21	22 Mar 43	St Nazaire	Flt Sgt J Murray	66	22 Oct 43	Kassel	Flt Sgt W Watson
22	26 Mar 43	Duisburg	Flt Sgt J Murray	67	3 Nov 43	Dusseldorf	Wnt Off H Carter
23	27 Mar 43	Berlin	Flt Sgt J Murray	68	18 Nov 43	Berlin	Wnt Off H Carter
24	29 Mar 43	Berlin	Flt Sgt J Murray	69	22 Nov 43	Berlin	Flt Sgt R Douglas
25	3 Apr 43	Essen	Sgt P Coldham	70	23 Nov 43	Berlin	Flt Sgt R Douglas
26	4 Apr 43	Kiel	Sgt P Coldham	71	26 Nov 43	Berlin	Wnt Off H Carter
27	9 Apr 43	Duisburg	Flt Sgt J Murray	72	2 Dec 43	Berlin	Flt Sgt K Goodwin
28	10 Apr 43	Frankfurt	Sgt J Williams	73	3 Dec 43	Leipzig	Flt Sgt R Douglas
29	13 Apr 43	Spezia	Flt Sgt J Murray	74	16 Dec 43	Berlin	Plt Off H Carter
30	16 Apr 43	Koblenz	Flt Sgt J Murray	75	20 Dec 43	Frankfurt	Plt Off H Carter
31	18 Apr 43	Spezia	Flt Sgt J Murray	76	23 Dec 43	Berlin	Plt Off H Carter
32	20 Apr 43	Stettin	Flt Sgt J Murray	77	29 Dec 43	Berlin	Flt Lt A Wales
33	27 Apr 43	Duisburg	Flt Sgt W Rose	78	1 Jan 44	Berlin	Plt Off J Howell
34	4 May 43	Dortmund	Flg Off J Henderson	79	5 Jan 44	Stettin	Plt Off J Hills
35	27 May 43	Essen	Sgt D J Strath	80	19 Feb 44	Leipzig	Flt Sgt J McCleery
36	29 May 43	Wuppertal	Sgt D J Strath	81	20 Feb 44	Stuttgart	Flg Off T Leggett
37	11 Jun 43	Dusseldorf	Flg Off J Henderson	82	24 Feb 44	Schweinfurt	Flg Off T Leggett
38	12 Jun 43	Bochum	Flg Off J Henderson	83	25 Feb 43	Ausberg	Flg Off T Leggett
39	14 Jun 43	Oberhausen	Flg Off J Henderson	84	18 Mar 44	Frankfurt	Plt Off K Morgan
40	16 Jun 43	Cologne	Flg Off J Henderson	85	22 Mar 44	Frankfurt	Plt Off K Morgan
41	21 Jun 43	Krefeld	Flg Off J Henderson	86	26 Mar 44	Essen	Flt Sgt R Allen
42	22 Jun 43	Mulheim	Flg Off J Henderson	87	30 Mar 44	Nuremburg	Flt Sgt V Neal
43	24 Jun 43	Wuppertal	Flg Off J Henderson	88	9 Apr 44	Villeneuve St George	Flg Off J Critchley
44	25 Jun 43	Gelsenkirchen	Flg Off J Henderson	89	10 Apr 44	Auloyne	Flg Off J Critchley
45	28 Jun 43	Cologne	Flg Off J Henderson	90	20 Apr 44	Cologne	Flg Off J Critchley

460 (RAAF) Squadron's G for George receiving a visit from the Australian Prime Minister, Mr John Curtin. (via Neil Mackenzie)

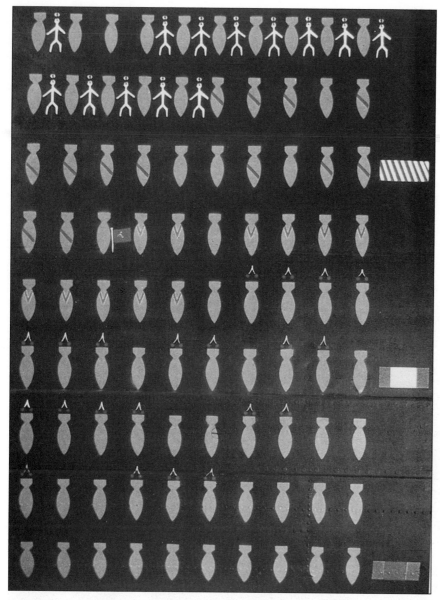

G for George's missions tally. The 'Saint' figures above some of the early ones indicate the missions flown with Flt Sgt J A Saint-Smith in command; those with cherries reflect the nickname of Flt Sgt (later Wnt Off and Plt Off) H A Carter.

tralia and New Zealand on fund raising and morale boosting trips and proved to be a popular attraction. Queenie achieved some notoriety in October 1943 when it became the largest aircraft to fly under the Sydney Harbour Bridge.

Upon its withdrawal from service, George was presented to Australia by Britain after logging 664 flying hours, at which time the decision was made to bring the aircraft to Australia to take over Queenie's promotional duties. George was an entirely appropriate choice because here was a Lancaster which not only had an extensive operational history behind it but one which had also served with an RAAF squadron.

George arrived at RAAF Amberley, Queensland, on 8 November 1944 after a four week flight from Prestwick via Montreal, San Francisco, Hawaii, Fiji and New Caledonia. The captain was Flt Lt E A Hudson DFC and Bar.

After completing its morale boosting and fundraising activities, George was declared surplus in July 1945 and flown to Canberra for preservation. Ten years later, after being stored outside and subject to the effects of weather and vandalism, the Lancaster was placed in the Australian War Memorial's Aeroplane Hall in Canberra, where it stands today, restored and resplendent in its 460 Squadron 'AR-G' markings. Queenie was not so fortunate; it was scrapped in 1948.

The following table lists all of G for George's 90 combat missions, noting the date, target and skipper.

G for George at the time of its arrival in Australia during November 1944, still carrying the British serial W4783. (via Neil Mackenzie)

(above) G for George at Laverton, near Melbourne in March 1945, by now wearing the RAAF serial A66-2. (via Neil Mackenzie)

(below) The first Lancaster to arrive in Australia, ED930/A66-1 'Queenie VI', in company with an RAAF Spitfire VC. (Defence PR)

BOEING B-29
SUPERFORTRESS

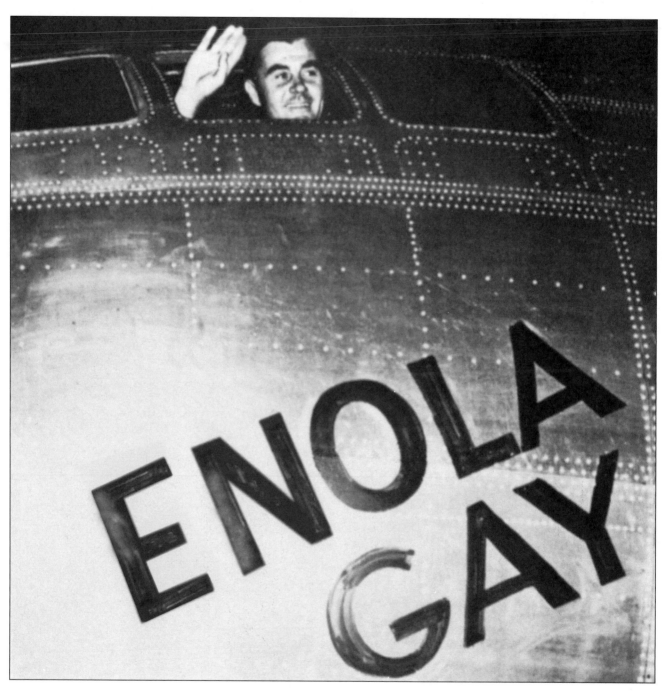

The most famous B-29 pilot in the most famous B-29. Paul Tibbets and 44-86292 'Enola Gay' before departure on the world's first atomic bomb raid on 6 August 1945.

BOEING B-29 SUPERFORTRESS

The Boeing B-29 Superfortress was a technological marvel of its time – a very large aeroplane with a combination of features which removed it from the standards established by 'conventional' heavy bombers of World War II such as the Lancaster and B-17 Flying Fortress.

Nearly twice as heavy at maximum weight and with a 36 per cent greater wing span than the B-17, the B-29 represented a new generation with tricycle undercarriage, pressurisation, remotely controlled defensive guns, numerous other advanced technical features and unrivalled payload/range characteristics. The B-29's contribution to Boeing's postwar expertise as a builder of large civil and military aircraft is another area which should not be underestimated.

The B-29 suffered early problems, particularly with its engines and partially as a result of very rapid development. The first prototype flew in September 1942 and the earliest production examples were coming off the production line just 12 months later, a remarkable achievement in itself considering the complexity and innovative nature of the aircraft.

As will be discussed later, it took some time for the Superfortress to become fully established in operational service, but when it did, the results were devastating. The best known B-29 missions are obviously the two which dropped atomic bombs on Hiroshima and Nagasaki in August 1945, thus bringing the Pacific War to a swift end. These missions represent the only times in the history of human conflict that atomic weapons have been used in anger.

Equally important in the history of the B-29 are the conventional bombing missions the aircraft flew against Japan in the final year of the war. Flying from the Marianas Islands, the Superfortresses embarked on a campaign intended to obliterate Japan's industries and cities. The latter – 'fire raids' using incendiary bombs – were particularly devastating and one of these missions against Tokyo in March 1945 is regarded as the single most destructive air raid ever, even more so than the atomic bomb raids.

The end of World War II did not mean the end of the B-29's operational career, the Korean conflict seeing to that and adding another chapter to the aircraft's history between 1950 and 1953.

From the B-29 was developed the heavier and more powerful B-50 Superfortress, which was built in comparatively modest numbers postwar. Between them the two 'Superforts' spawned numerous subvariants, many of them associated with the then developing art of aerial refuelling, another area in which the Boeing aircraft would become intimately associated in the 1950s and beyond.

B-29 production amounted to 3,960 aircraft between the first flight in September 1942 and final rollout in May 1946. The vast majority of these were built in 1944 and 1945 at four factories throughout the USA, a remarkable achievement of mass production of a large and complex aircraft and indicative of the USA's industrial might. To the B-29 production tally should be added the 370 B-50s built between 1947 and 1953.

Towards The Big Bomber

By the late 1930s Boeing had developed considerable expertise in the building of large aircraft, among them the YB-9 bomber of 1931, the Model 247 ten passenger airliner of 1933, the B-17 Flying Fortress, the Model 314 Clipper flying boat and the huge XB-15 experimental long range bomber.

The XB-15 was developed in parallel with the B-17 and came about as

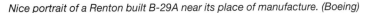

Nice portrait of a Renton built B-29A near its place of manufacture. (Boeing)

A B-29A displays some of the Superfortress's salient features including the position of the two bomb bays and the five gun turrets. (Boeing)

(top and bottom) Two views of Boeing's first 'big bomber', the mammoth, one-off XB-15 of 1937, at the time the largest and heaviest aircraft built in the USA. Wing span was 149 feet (45.4m) and the aircraft was capable of an endurance of over 24 hours. (Boeing)

a result of a 1934 US Army Air Corps requirement for a long range bomber capable of carrying a 2,000lb (907kg) bomb load over the then unheard of distance of 5,000 miles (8,000km). As it turned out, the specification was fanciful because it exceeded the available technology of the time, but for Boeing it was invaluable experience which in combination with the work associated with the B-17 stood it in good stead for the future.

Boeing submitted its Model 294 to meet the specification, at that time the largest and heaviest aircraft ever built in the USA with a wing span of 149 feet (45.4m) and an all up weight of over 70,000lb (31,750kg). A single Model 294 was built, first flying in October 1937. It was at first given the military designation XBLR-1 (for 'experimental bomber long range type one') but this was subsequently changed to XB-15. Its military serial number was 35-277.

The XB-15 had originally been designed around four 1,000hp (750kW) Allison V-3420 water cooled vee-configuration powerplants but their unavailability led to the fitting of 850hp

The first Model 307 Stratoliner (NX 19901) recorded its first flight on 31 December 1938. This advanced pressurised airliner incorporated the wings, tail surfaces, engines and undercarriage of the B-17C Fortress with a new fuselage capable of carrying 33 passengers. Note the Boeing 247 and Douglas DC-3 on the ground. (Boeing)

(635kW) Pratt & Whitney R-1830 Twin Wasp radials instead. These left the XB-15 grossly underpowered and performance suffered as a result, the aircraft managing a top speed of just 197mph (317km/h) and a service ceiling of only 18,900ft (5,760m). Its range performance was quite remarkable for the era, however, a maximum (without payload) of 5,130 miles (8,255km) being achieved while with a 2,500lb (1,134kg) bomb load aboard it could travel 3,400 miles (5,470km). Maximum bomb load was 8,000lb (3,630kg).

Of all metal construction, the XB-15 featured a streamlined fuselage with a slim rear fuselage design and stream-lined engine cowlings. Everything about it was big including the double wheeled main undercarriage and the wing area of 2,780sq ft (258.2m^2), or nearly twice that of the B-17.

Some interesting features were incorporated, much of them stemming from the aircraft's very long endurance with missions of 24 hours or more possible. A crew of ten was needed and the XB-15 was fitted with bunks and cooking facilities to cater for those who were off duty at a given time. Duties were arranged much like a ship, with 'watches'.

The XB-15's very thick wing inboard wing sections allowed space for a tunnel behind the engine na-celles so the flight engineer could get to the engines in flight and electrical power was provided by two petrol powered generators.

The lack of a suitably powerful engine was the main reason for the XB-15's failure and orders for a developed version, the Y1B-20, were cancelled. The sole XB-15's career didn't quite end there. It was used for research flying and then served well into the 1940s as a cargo transport under the new designation XC-105.

Superfortress Genesis

The failure of the XB-15 did not deter Boeing from continuing to study long range bombers, work that

A Stratoliner with the modified tail surfaces applied to all aircraft after the loss of the prototype due to structural failure during an inadvertent spin. Production of this advanced airliner amounted to only 10 aircraft up to 1940 but it nevertheless provided Boeing with part of the bank of knowledge it was building up on large aeroplanes incorporating advanced systems. These lessons would soon be applied to the B-29. (Boeing)

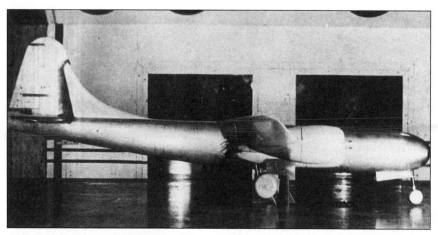

Wind tunnel model of the Boeing Model 345, which would emerge as the B-29 Superfortress.

would shortly be put to good use when the requirement which eventually resulted in the B-29 was issued. This work was carried out against a background of official disinterest caused by a lack of funding, a government policy which saw the Army Air Corps as a mainly defensive entity and opposition to Air Corps expansion from the US Navy. As discussed in the B-17 section of this book, the Navy jealously guarded its 'right' to control the seas and any attempt by the Air Corps to usurp some of this right was strenuously opposed. As late as 1938 Air Corps aircraft were not supposed to venture more than 100 miles (161km) out to sea.

Boeing, in the meantime, continued its studies on long range bombers throughout the late 1930s, projects examined including the Model 322, a pressurised and tricycle undercarriage development of the B-17 not dissimilar to the Model 307 Stratoliner commercial transport; the Model 333 with tandem pairs of Allison V-1710 liquid cooled engines buried in a very thick section wing; the 334 similar but with Pratt & Whitney radial engines and twin fins and rudders; and the 334A with conventionally mounted Wright R-3350 radial engines and single fin and rudder. Bearing a close resemblance to the forthcoming B-29, a mockup of this aircraft was constructed in 1939.

Another project was the Model 341 of 1939 with a high aspect ratio wing incorporating a new high lift aerofoil. Power was to be provided by four 2,000hp (1,490kW) Pratt & Whitney R-2800 radials and performance estimates included a maximum speed of 405mph (651km/h) and a range of no less than 7,000 miles (11,265km) carrying one ton (1,016kg) of bombs. As will be seen, this aircraft seemed tailor made for the very long range bomber specification when it was finally issued.

Despite the political situation of

the time, the Army Air Corps and the concept of the long range bomber had some friends. Among them was Maj Gen Henry H ('Hap') Arnold, chief of the Air Corps from 1937 and famed aviator Charles Lindbergh who made a study tour of German aircraft factories and *Luftwaffe* bases. He reported Germany's advanced state of preparation, both in the military and industrial senses and the fact that in his opinion, America was being left behind.

Lindberg's reports prompted Maj Gen Arnold to commission Brigadier General Walter Kilner to head a committee which would examine and report on the long term needs of the Air Corps. The Kilner Report was delivered on 1 September 1939 – the day Germany invaded Poland – and among its recommendations was the immediate initiation of the development of new long range medium and heavy bombers. In one sense, the publication of the report could not have been more timely because the events in Europe prompted many who had previously been at best luke warm about long range bombers to think again. Although the USA was officially neutral, growing numbers of senior military men (and politicians) were thinking that somewhere along the line their country would inevitably become involved.

An important early convert to the cause was the Army Chief of Staff, Gen Malin Craig, who supported Arnold on the establishment of the Kilner Committee. Craig retired on the same day the Kilner Report was delivered, but his successor, Gen George C Marshall, was even more enthusiastic about the future of the

A very big bomber from Boeing's competitor. The Douglas XB-19 (pictured) was designed to meet the same requirement as the Boeing XB-15 but was even larger, spanning a massive 212 feet (64.6m) and designed for a maximum weight of 780,000lb (353,800kg), not much less than the Boeing 747-200! Powered by the same Wright R-3350 engines as the B-29, the sole XB-19 first flew three years behind schedule in 1941.

Army Air Corps and long range bombers. His influence was fundamental to the Air Corps (restyled as the US Army Air Force from June 1941) developing the way it did.

The Boeing Model 345

The chain of events which resulted in the B-29 developed quickly from late 1939. In November, Gen Arnold requested and received authority (as well as the all important budgetary commitment) to approach aircraft manufacturers with a specification for a very long range bomber. The specification was prepared by a team under the command of AAC Materiel Command's Capt Donald L Putt, a test pilot who had been involved in the B-17 programme.

Request for Data R-40B and Specification XC-218 were formally issued to Boeing, Lockheed, Douglas and Consolidated-Vultee in late January 1940. It called for a four engined bomber capable of a maximum speed of 400mph (644km/h) and a range of 5,333 miles (8,580km) carrying one ton (1,016kg) of bombs.

Consolidated responded with its Model 33, Lockheed with the Model 51-81-01, Douglas with the Model 332F and Boeing with the Model 341. They were allocated the Air Corps designations XB-32, XB-30, XB-31 and XB-29, respectively.

Lockheed and Douglas were early withdrawals from the competition and in May 1940 Boeing and Consolidated were announced joint winners. Changes to the requirement resulting from experience gained in European combat (self sealing fuel tanks, increased armour protection and more defensive firepower) resulted in Boeing proposing the Model 345, basically an enlarged 341 with Wright R-3350 18 cylinder radial engines replacing the original Pratt & Whitneys.

Boeing's design was technically streets ahead of Consolidated's offering with a myriad of advanced features. Although these manufacturers' designs were the only two which

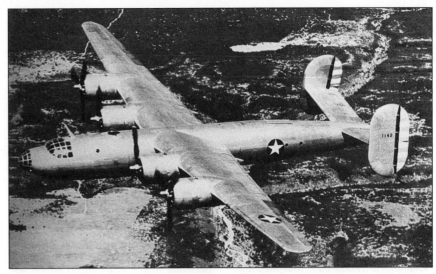

The contest which resulted in the B-29 Superfortress entering production also saw the Consolidated XB-32 built. This is the second prototype; the small number of production aircraft which were built had single fin and rudder assemblies.

reached the hardware stage, B-32 Dominator production was limited to just a dozen examples.

The Boeing 345/XB-29's innovations included pressurisation (the first for a purely military application), tricycle undercarriage (the first for a heavy bomber) and remotely controlled defensive armament. Power was to be provided by four 2,200hp (1,640kW) R-3350 turbo-supercharged engines and a maximum bomb load of 16,000lb (7,257kg) could be carried in two bomb bays fore and aft of the wing. It was a big aircraft, spanning 141ft 3in (43.05m) and having a design maximum weight of (initially) well over 100,000lb (45,360kg). The prototype and evaluation batch B-29s were driven by three bladed propellers; production aircraft had Hamilton Standard four bladers of 16ft 7in (5.06m) diameter.

The weight of the aircraft created its own problems, the result of a very high wing loading which was nearly twice that of a B-17E Fortress. In order to keep approach speeds down to an acceptable level, Boeing's new and very efficient high aspect ratio (11.5

to 1) wing was fitted with large Fowler flaps which added 20 per cent to the wing area when they were lowered.

Incorporating pressurisation introduced new problems, the main one revolving around how to deal with the opening of the bomb bay doors whilst maintaining pressurisation in the crew compartments. The solution was found in dividing the aircraft into two separate pressurised compartments, one incorporating the nose section which housed two pilots, the bombardier, flight engineer, navigator and radio operator and the other in a mid-rear fuselage section which accommodated the central fire control gunner, two side gunners and the radar operator.

These two compartments were joined by a 40 feet (12m) long tunnel of 34 inches (86cm) diameter which passed over the forward bomb bay, wing spar and rear bomb bay. The tunnel was just about large enough for a man to crawl through – with some difficulty when in full flying gear. The rear pressure bulkhead was located below the leading edge of the fin but the tail gunner (housed in a com-

The first XB-29 Superfortress (41-002) in its olive drab finish. This aircraft recorded its maiden flight on 21 September 1942.

The first XB-29 in flight. Early test flights quickly revealed serious problems with the Wright R-3350 engines.

partment another 40 or so feet (12m) away from the central compartment) had his section pressurised via two small pipes through the rear fuselage.

The B-29's pressurisation system enabled an 8,000ft (2,400m) cabin altitude to be maintained while the aircraft was flying at 30,000ft (9,144m).

The B-29's remote controlled gunnery system was a major innovation.

There were five gun positions: upper forward, upper aft, lower forward, lower aft and tail. Three of the four fuselage mounted positions had a pair of 0.50in Browning machine guns in the turrets, the exception being the forward upper which had three guns. The tail turret initially featured two machine guns and a 20mm cannon.

Four companies (Bendix, General

Electric, Westinghouse and Sperry) developed gunnery systems for the B-29 and bid for the contract. Sperry was the original winner with its retractable turret/periscope sighting system but this was changed to General Electric's computer controlled system by the time the aircraft entered production. The decision to change was made not because of any problem with the Sperry equipment but because GE's system was much more advanced and theoretically more effective.

Overall control of the defensive armament was performed by the central fire control gunner, sitting in the mid fuselage position and looking through a perspex blister on the top of the fuselage. He was flanked by two other gunners seated in adjacent large blisters in the fuselage side. The bombardier in the nose compartment also doubled as a gunner while the tail gunner had his own compartment in the extreme rear of the aircraft.

B-29 fuselages under construction clearly show the circular cross section and pressurised tunnel which passed above the bomb bays and joined the forward and aft crew stations. (Boeing)

The complex system allowed the central fire control gunner to assign turrets to specific gunners, depending on their view of the action. Each gunner had a primary weapon he looked after, but the system allowed control of any turret to be given to any gunner (except the tail man), and any gunner (again, apart from 'tail end Charlie') could operate two turrets at a time if necessary. Sighting was through a device at the blister windows which rotated the turrets and raised and lowered the guns as it was moved. The computer allowed for smooth tracking as well as compensating for lead angle, distance, altitude and temperature.

These and the other systems installed on the B-29 resulted in a huge demand for electrical power, necessitating the installation of a large number of especially designed generators in the aircraft.

The B-29's two bomb bays were each about 14 feet (4.25m) long, each capable of carrying up to two 4,000lb (1,814kg) weapons or a mixture of smaller bombs. Another innovation was the incorporation of the 'Intervalometer', a device which regulated the dropping of bombs alternatively from the two bomb bays so as to maintain centre of gravity.

Preparation For Production

Boeing received an order for two XB-29 flying prototypes and a static test airframe on 24 August 1940 in a contract worth $US3.615m. The company had completed full scale mockups by the following November and a third XB-29 was ordered in December 1940. These aircraft were the only B-29s built at Boeing's Seattle facility and were allocated the serial numbers 41-002, 41-003 and 41-18335.

Even though first flight was still some 21 months away at this stage, production orders were quickly placed and the organisation under which they would be built was established. The first order was placed in June 1941 for 14 YB-29 service evaluation aircraft followed by an order for 250 production aircraft in September. The Japanese raid on Pearl Harbour and America's resulting involvement in World War II provoked further orders. By the time the first XB-29 was flown, Boeing had contracts covering 1,849 production aircraft.

The initial production order for 250 B-29s prompted Boeing to build a new facility at Wichita, Kansas, while subsequent orders during 1942 and beyond resulted in the establishment of additional production lines by Boeing (Renton, Washington), Martin (Omaha, Nebraska) and Bell (Atlanta, Georgia). The result was an illustration of America's industrial might as

The third XB-29 prototype (41-18335) first flew in June 1943.

the facilities for mass production of a very large and complex aircraft were organised well in advance.

Flying Into Trouble

The first XB-29 (41-002) was rolled out of the Seattle factory in early September 1942, Boeing's director of aerodynamics and flight research, Edmund T ('Eddie') Allen and chief test pilot Al Reed conducting taxying tests at Boeing Field, some of the high speed ones resulting in 'hops' to a height of about 15 feet.

The XB-29's real first flight took place on 21 September 1942 with

The B-29 flightdeck, photographed from behind the copilot's seat. This area was nicknamed 'The Greenhouse' for obvious reasons and the muilti-paned bullet proof glass provided excellent visibility.

Allen in command and Reed acting as copilot. The flight lasted 75 minutes and revealed no serious snags, although Allen had previously noted the Wright engines' tendency to overheat after a short period of time. These were prophetic observations.

Donald Putt, the man who had put together the specification which led to the B-29, flew the prototype the next day and therefore became the first US Army Air Force pilot to do so. He enthused about the aircraft, saying it was "easier to fly than a B-17" and it was "unbelievable for such a large plane to be so easy on the controls". Putt also commented on the XB-29's speed ("faster than any previous bomber"), although it's interesting to note that its maximum speed was short of the required 400mph (644km/h), about the only area where it failed to meet the specification.

The euphoria associated with these successful early flights soon disappeared as problems began to appear. Most were associated with the B-29's complicated and underdeveloped engines. The prototype logged only 27 flying hours between first flight and December 1942, suffering numerous engine failures and problems with the constant speed propeller operating mechanism resulting in runaway engines. Aerodynamically, Boeing's engineers had got it right, the only significant change being removal of the rudder boost.

The first XB-29 survived an inflight engine fire on 28 December 1942 and the maiden flight of the second prototype (41-003) two days later resulted in a similar but more serious emergency. On that occasion, the No 4 (starboard outer) engine caught fire just six minutes into the flight. Eddie Allen was in command and despite thick smoke in the cockpit he was able to bring the aircraft safely back to earth where ground crews extinguished the blaze.

At this stage it appeared the whole B-29 programme was in serious trouble as no amount of investigation seemed to be able to solve the problems. Unscheduled engine changes became standard procedure and more significantly, many were having doubts about the B-29 generally.

Seattle Tragedy

Unfortunately, things got much worse before they improved.

On 18 February 1943, Eddie Allen and his ten man crew departed Boeing Field on a test flight in the second XB-29. Eight minutes after takeoff the No 1 engine caught fire. It was shut down and the propeller feathered. Allen decided to return to land, thinking the fire was out. Unfortunately, it was not and undetected by the crew was in fact eating through the front spar of the port wing. It wasn't until the aircraft was on base leg that the fire was noticed, Allen attempting to get the aircraft down as quickly as possible as it was obvious by the intensity of the fire that it couldn't survive for long.

Allen lost control at a height of about 250 feet (76m), the XB-29 crashing into a meat packing plant about three miles (5km) from the runway, killing all on board and 19 on the ground.

The investigation into the accident laid most of the blame at the door of the engine manufacturer, Wright, although the USAAF was itself criticised for putting Wright into a position where quick results were being demanded of it.

Engine problems continued to dog the B-29 well into 1944 before they were finally solved. It was found that excessive cylinder head temperatures resulted from poorly lubricated valves and that a faulty fuel induction system design was responsible for the fires. The latter wasn't positively proven until 1944 when the first XB-29 suffered a fire, but quick extinguishing of it meant the engine could be examined after landing. Previously, the evidence had been destroyed by the fire itself.

Reorganisation

The Seattle crash brought the B-29 test programme virtually to a stop. The political ramifications were potentially serious, as the very long range bombing aspirations of the USAAF in general and 'Hap' Arnold in particular could well have disappeared in the fireball which accompanied the accident.

Arnold arranged a complete reorganisation of the B-29 programme in order to ensure its survival. A new management programme was established under the command of Brig

The B-29's forward upper gun turret, showing the twin 50-calibre machine guns and the gunner's station. The aircraft's computerised gun control system allowed any gunner (except 'Tail End Charlie') to control any two turrets apart from the tail installation. (Boeing)

Gen Kenneth B Wolfe with the title 'B-29 Special Project'. Wolfe and his team would take charge of the entire project, not just the flight testing and engineering aspects of it but crew selection and training as well. This was a new and radical approach for the military, and it worked. Wolfe's team was up and running by mid April 1943, their primary aim being to have the B-29 ready for combat by the end of the year. This target was just about met and the B-29 carried on from there.

Despite this, another near disaster which surely would have ended the B-29 programme had it occurred was to be faced. The third XB-29 (41-18335) was first flown on 16 June 1943 but it nearly had a very short life as it was discovered shortly before a planned takeoff of few days earlier that the aileron cables had been connected in reverse. The loss of that aircraft under those circumstances would have put enormous pressure on the project.

The first of 14 service evaluation YB-29s (41-36954) flew from Boeing's new Wichita plant on 27 June 1943. This and subsequent variants are discussed in the following chapter.

(above) B-29As in mass production at Boeing's Renton plant. The ability of American industry to produce over 3,900 advanced and complex Superfortresses mostly in the space of less than two years, speaks for itself. (Boeing)

(below) Boeing Wichita's 1,000th B-29 on February 1945 in company with the 10,346th Stearman Kaydet trainer. The aircraft are covered with money as part of the 'March of Dimes' programme to raise funds for polio research. Contributions from employees at the factory reached $US10,562. (Boeing)

(top) 42-93888 is a B-29A. She is pictured here in front of Lake Washington at Boeing's Renton plant. (bottom) The Guppy conversion of the Boeing 377/C-97 was originally designed to ferry rocket booster stages for the American space programme. (Boeing/Gordon Reid)

Two survivors in the hands of the world famous Pima County Air Museum, near Tuscon Arizona, are (top) this B-29 'City of Philadelphia' and (bottom) a KB-50. (Keith Myers)

YB-29 SUPERFORTRESS

The first of 14 YB-29 Superfortress service test and training aircraft (41-36954) was flown on 27 June 1943, the predecessor of 1,634 B-29s which would eventually emerge from Boeing's new factory at Wichita, Kansas.

Powered, like the prototypes, by the same 2,200hp (1,640kW) R-3350 Cyclone engines with General Electric B-11 turbosuperchargers driving three bladed propellers of 17ft 0in (5.18m) diameter, these Superfortresses were allocated to the Accelerated Test Service Branch (ATSB), originally based at Wright Field, Ohio, and then Smokey Hill Field near Salina, Kansas. The ASTB was established as a result of the Wolfe Special B-29 Project discussed in the previous chapter, which had been given top priority in order to get the aircraft developed and into service, and crews trained.

B-29 SUPERFORTRESS

The 'standard' B-29 was built at Boeing's new Wichita plant (1,620) by Martin at Omaha (536) and by Bell at Atlanta (357) bringing total production of this model to 2,513 aircraft. The first Boeing example was flown in June 1943 and the last in October 1945; Martin's aircraft came off the

The first service evaluation YB-29 41-36954 later in its service life when it had been converted to XB-39 configuration with four Allison V-3420 liquid cooled inline engines in place of the usual Wright R-3350 radials. (Boeing)

line between January 1944 and September 1945 and Bell's between February 1944 and January 1945.

Compared to the prototypes and evaluation Superfortresses, production B-29s featured four bladed Hamilton Standard propellers of 16ft 7in (5.05m) diameter, although a few late aircraft were fitted with reversible Curtiss propellers with electrical rather than hydraulic pitch change actuation.

The powerplant was initially the R-3350-23 of the service test aircraft but ongoing development to solve the engine's problems resulted in different versions of the engine being installed as production progressed. The R-3350-41 engine with improved

cooling was introduced midway through the production run and later aircraft were fitted with further improved -57 and -57A engines. R-3350-29, and -59 engines were also installed in some cases, the different versions incorporating mainly minor modifications. In all cases the engines were rated at 2,200hp (1,640kW) for take-off, 2,000hp (1,490kW) maximum continuous and 2,300hp (1,715kW) war emergency power.

It took some time for the B-29's engine problems to be resolved and development in this area concentrated on reducing and eventually removing the risk of fire. In the meantime, the cost was high with at least 19 aircraft lost due to this cause between February 1943 and September 1944. It wasn't until then that the basic problems – excessive cylinder head temperatures due to improperly lubricated valves and faulty fuel induction systems – were found.

Defensive armament also varied as production progressed. The initial fit was a pair of 0.50in machine guns in each of the four remotely controlled turrets plus two 'fifties' and a 20mm cannon in the tail. The need for greater frontal protection resulted in the forward upper turret being fitted with four guns during the production run, while in late aircraft the 20mm cannon was deleted from the tail position as accurate aiming was found to be difficult due to the very different trajectory characteristics of the cannon and machine gun rounds.

The maximum bomb load which could be carried was 20,000lb (9,072kg) comprising various combinations of high explosive and incendiary weapons. The bomb bay doors were initially hydraulically operated with a seven seconds' activation time but from late 1944 all B-29s were fitted with pneumatic doors which could snap shut in less than one second.

The B-29's initial fuel capacity was

A view of B-29 44-87775, highlighting the Superfortress's efficient, high aspect ratio wing.

Production of the standard B-29 was undertaken by Boeing, Martin and Bell, and 2,513 of this variant were manufactured between them. 44-87775 is a product of Boeing's Wichita factory.

5,470 USgal (29,706 l) in four wing tanks, this increasing to 6,803 USgal (25,752 l) with the addition of a centre section tank early in the production run. Semi permanent tanks in the bomb bays could increase the B-29's maximum fuel capacity to 9,363 USgal (35,443 l) giving a ferry range of up to 6,000 miles (9,655km).

The standard B-29 crew complement was 11 comprising pilot, copilot, bombardier, navigator, flight engineer, radio operator, radar operator, central fire control gunner, port and starboard gunners and tail gunner. The fitting of increasingly sophisticated radar and electronic countermeasures (ECM) equipment sometimes warranted the addition of two extra crewmembers who were specialists in these fields.

Early B-29s were fitted with AN/APN-4 LORAN (long range) constant beam navigation aids, this being replaced by the more capable AN/APN-9 system later on. Another important item of equipment was the installation of AN/APQ-13 bombing/navigational radar with a 30in (76cm) antenna in a retractable underfuselage radome between the bomb

bays. This radar was based on the British H2S system and provided a rough radar image of the ground. Improved AN/APQ-7 'Eagle' radar was later fitted in an aerofoil shaped radome under the fuselage.

In 1945, the average cost of a B-29 was put at $US509,465 or 2.7 times that of a B-17G which cost $US187,742.

Production Delays

The large number of orders placed for the B-29 even before its first flight had ensured the establishment of

four production facilities for the aircraft by the beginning of 1944 – Boeing Wichita, Martin Omaha, Bell Atlanta and Boeing Renton, the latter producing only the B-29A variant described below. Despite this, the complexity of the aircraft, associated problems and an ambitious production schedule contributed to delays in the aircraft reaching service.

There was considerable political pressure to get the B-29 into operational service, pressure which resulted from the January 1943 Casablanca conference where US President

A B-29 with the partially retracted radome for the AN/APQ-13 radar visible under the lower centre fuselage.

A late production B-29 (45-21783) photographed postwar while being used for cosmic ray research duties on behalf of the National Geographic Society.

quickly as possible. Some things can't be rushed too much, even in wartime, and making an advanced and even revolutionary aircraft such as the B-29 ready for combat deployment is one of them.

There were the engine problems, problems with the complex gunnery system, problems with the radar, problems with other systems and components, all of which took time to rectify. By the middle of January 1944 about 100 B-29s had been flown of which only 16 were airworthy and none had been issued to the 58th Bombardment Wing, the initial operational recipient of the aircraft.

Franklin Roosevelt promised China's Generalissimo Chaing Kai-shek considerable aerial support to help prevent Japan from overrunning China. Three hundred bombers were promised (not necessarily B-29s) but it soon became obvious that the new bomber would be the only one capable of performing the difficult and long range missions in that part of the world. By the time of the Quebec conference in August 1943 Roosevelt had promised 200 B-29s to help

China by the following March. At that point, only the prototypes and pre production aircraft had flown. The first production aircraft would not appear until September 1943.

A massive production effort saw 200 B-29s in the air by the end of February 1944 in a programme inspired and pushed by General 'Hap' Arnold. Called 'The Battle of Kansas', this programme was basically an 'all hands on deck' affair intended to make the B-29 combat worthy as

Another major problem was that of training crews, exacerbated by a decision to provide two crews for each aircraft due to the long endurance missions the B-29 would fly. It wasn't simply a matter of teaching pilots how to fly the aircraft, there was also the need for specialist training for radar operators and gunners as the B-29 introduced new techniques which could be taught to some extent in simulators (and this was done) but experience in the real thing was

B-29s in production, in this case at Boeing's Wichita facility. The two Boeing factories at Wichita and Renton between them turned out 100 Superfortresses per month at their peak in 1945. (Boeing)

the best way to learn. The pilots and navigators, too, needed to know how to deal with the challenges created by the B-29's capabilities of ranges which were longer and speeds which were greater than existing bombers.

The first operational B-29s were scheduled to deploy to India in mid March 1944 and shortly afterwards to China from where they would fly against the Japanese in the China-Burma-India (CBI) theatre. The aircraft were nowhere near ready by the due date, prompting the 'Battle of Kansas' (or 'Kansas Blitz' as it was also known) effort. Around the clock work by all concerned saw 150 B-29s ready for service by the middle of April 1944.

The 58th Bombardment Group's B-29s began ferrying to the CBI in the second half of April, although problems immediately began to haunt what would turn out to be an ill-starred adventure. This will be discussed in a following chapter, but five B-29s crashed within a week of each other near Karachi during the first deployment flights. Overheating engines was the cause – exhaust valves in the rear bank of cylinders were melting from the heat – and the B-29s were grounded for a fortnight while this latest setback was sorted out.

For the record, the B-29's first operational mission took place on 5 June 1944 when 98 aircraft departed India to raid the Makasan railway yards at Bangkok. It was not an auspicious start to the Superfort's career. Few bombs fell on the target, 14 aircraft aborted before reaching it, five crashed on landing and no fewer than 42 were forced to divert to other airfields due to a lack of fuel.

B-29A SUPERFORTRESS

Built exclusively at Boeing's Renton plant, the B-29A differed from the standard aircraft mainly in the method of wing construction. The normal B-29 wing was manufactured in six major sub assemblies with the single piece centre section integral with the fuselage. On the B-29A, a two piece centre section was assembled as a single unit and then attached to the airframe. A slight reduction in fuel capacity resulted, as did an increase in wing span of one foot (30cm) to 142ft 3in (43.36m). Basic empty weight went up by 1,220lb (553kg) to 71,360lb (32,369kg).

B-29 SUPERFORTRESS

Powerplants: Four Wright R-3350-23, -41 or -57/A Cyclone 18 cylinder two row radial engines each with two General Electric B-11 exhaust driven turbosuperchargers, rated at 2,300hp (1,715kW) war emergency power, 2,200hp (1,640kW) for takeoff, 2,200hp (1,640kW) at 25,000ft and 2,000hp (1,490kW) max continuous; Hamilton Standard four bladed constant speed and feathering propellers of 16ft 7in (5.05m) diameter; normal fuel capacity 6,803 USgal (25,752 l) in wing and centre section tanks, provision for extra 2,560 US gal (9,690 l) tanks in bomb bay.

Dimensions: Wing span 141ft 3in (43.05m); length 99ft 0in (30.18m); height 27ft 9in (8.46m); wheel track 28ft 6in (8.68m); tailplane span 43ft 0in (13.10m); wing area 1,736sq ft (161.3m²).

Weights: Basic empty 70,140lb (31,816kg); maximum takeoff 138,000lb.

Armament: (Defensive) General Electric computerised remotely controlled turret system comprising upper forward, upper rear, lower forward and lower rear turrets each with two 0.50in machine guns with 500 or 1,000 rounds per gun, forward upper turret with four 0.50in guns in later aircraft (500 or 875 rpg); powered rear turret with two 0.50in machine guns (500 or 1,000 rpg) plus one 20mm cannon (100 rounds), deleted on later aircraft.

(Offensive) Two internal bomb bays capable of carrying a total load of 20,000lb (9,072kg). Typical loads include 40 500lb (227kg), four 4,000lb (1,814kg), eight 2,000lb (907kg), 12 1,600lb (726kg) or 12 1,000lb (454kg) bombs.

Performance: Max speed 266kt (493km/h) at 5,000ft, 299kt (553km/h) at 20,000ft, 310kt (575km/h) at 30,000ft; max cruise speed 286kt (529km/h) at 20,000ft; range cruise 200kt (370km/h); initial climb 900ft (274m)/min; time to 30,000ft (9,144m) 72min; service ceiling 31,850ft (9,700m); range with 5,000lb load 2,825nm (5,230km), with 12,000lb (5,440kg) load 2,300nm (4,260km), with 20,000lb load 1,695nm (3,138km); max ferry range (bomb bay tanks) 4,900nm (9,076km).

A B-29A Superfortress takes off from the airfield near Boeing's Renton factory. The B-29A differed from the standard model mainly in the method of wing construction. (Boeing)

(above) B-29As nearing completion at Renton. Production of this variant amounted to 1,119 aircraft, all of them built at Renton. (Boeing)

(right) The B-29A production line at Renton during 1944.

Boeing Renton turned out 1,119 B-29As from January 1944 and was the only factory to produce the aircraft after the surrender of Japan, the final aircraft rolling out on 28 May 1946. The first B-29A was serialled 42-93824 and the last (the final Superfortress built) 44-62328.

B-29B SUPERFORTRESS

With the B-29 raids over Japan attracting less fighter opposition than had been expected, the opportunity was taken to develop a new stripped down variant with the fuselage turrets and computerised gun control system deleted, leaving only the tail guns (normally three 0.50in machine guns) in place. The result was the B-29B, 311 of which were built by Bell at Atlanta between January and September 1945 starting with serial number 42-63704. Most were fitted with R-3350-41 engines.

The normal crew was reduced to seven or eight with the two fuselage gunners no longer necessary, although the central fire control gunner was sometimes carried as an observer and the bombardier's role could be taken over by the radar operator. Improved AN/APQ-7 bombing/navigational radar was usually carried.

Deletion of the gunnery control system reduced the empty weight of the B-29B by about 1,140lb (517kg) compared to the standard B-29 and by 2,360lb (1,070kg) by comparison with the B-29A to 69,000lb (31,298kg). This in conjunction with fewer crew members improved the aircraft's payload-range performance with the result that a 16,000lb (7,257kg) bomb load could be carried over a range of about 2,300nm (4,260km), representing a very useful 33 per cent increase in payload over a similar distance by comparison with a standard B-29.

Most B-29Bs were delivered to the USAAF's 315th Bomber Wing which was mainly involved in night missions against Japan's petroleum, oil and lubricants facilities.

SUPERFORT SUBVARIANTS

Several specialist role B-29 variants were produced by conversion of existing aircraft both during and after World War II, resulting in new designations, while other variants were planned and either not proceeded with or resulted in single prototype conversions only, often for trials use. Some of these are listed below:

One of 311 B-29Bs built by Bell at Atlanta in 1945. This aircraft (44-84084) is carrying AN/APQ-7 'Eagle' radar (with its 'wing' ventral radome just visible) and short range radar under the tail guns.

BOEING B-29 SUPERFORTRESS PRODUCTION SUMMARY

Notes: B-29 manufacturing plants were Boeing Seattle, Washington (prototypes only); Boeing Wichita, Kansas; Boeing Renton, Washington; Bell Atlanta (Marietta), Georgia; and Glenn L Martin Omaha, Nebraska.

The Wichita listing (USAAF Nos 42-6205 to 6454) includes 10 B-29s which were manufactured at that plant but assembled at Atlanta and Omaha (5 at each).

Model	Qty	US Serials	Factory	Remarks
XB-29	2	41-002/003	Seattle	41-002 ff 21/09/42
XB-29	1	41-18335	Seattle	ff 16/06/43
YB-29	14	41-36954/36967	Wichita	ff 27/06/43
B-29-BW	250	42-6205/6454	Wichita	first built 09/43
B-29-BW	500	42-24420/24919	Wichita	
B-29-BW	500	44-69655/70154	Wichita	
B-29-BW	200	44-87584/87783	Wichita	
B-29-BW	180	45-21693/21872	Wichita	last built 10/45
B-29-MO	112	42-65202/65313	Omaha	first built 01/44
B-29-MO	87	42-65315/65401	Omaha	
B-29-MO	100	44-27259/27358	Omaha	
B-29-MO	232	44-86242/86473	Omaha	last built 09/45

Model	Qty	US Serials	Factory	Remarks
B-29-BA	352	42-63352/63703	Atlanta	built 02/44-01/45
B-29B-BA	48	42-63704/63751	Atlanta	first built 01/45
B-29B-BA	263	44-83890/84152	Atlanta	last built 09/45
B-29A-BN	300	42-93824/94123	Renton	first built 01/44
B-29A-BN	819	44-61510/62328	Renton	last built 05/46

B-29 SUPERFORTRESS – PRODUCTION BY FACTORY

Note: The table reflects final assembly numbers, therefore 10 B-29s which were manufactured at Wichita but assembled at Omaha and Atlanta (5 each) appear in the latter tallies.

	XB-29	YB-29	B-29	B-29A	B-29B	Total
Boeing (Seattle)	3	–	–	–	–	3
Boeing (Wichita)	–	14	1620	–	–	1634
Boeing (Renton)	–	–	–	1119	–	1119
Martin (Omaha)	–	–	536	–	–	536
Bell (Atlanta)	–	–	357	–	311	668
Totals	3	14	2513	1119	311	3960

B-29C: Designation of proposed enhanced performance development of basic B-29 bomber with improved R-3350 engines installed. Project cancelled.

B-29D: Version of Superfortress powered by 3,500hp (2,610kW) Pratt & Whitney R-4360 engines and featuring substantially increased weights. Development and production undertaken post war as the B-50 Superfortress as described below.

F-13/RB-29: Photographic reconnaissance conversion, the first of which (based on early production Wichita built B-29 42-6412) was flown in 1944. A further 117 conversions were performed (30 at Wichita and 87 at Renton) under the designation F-13A, the first of them entering service in the Pacific theatre in October 1944.

Equipped with six K-18 and K-22 cameras in the bomb bays along with bomb bay fuel auxiliary fuel tanks, the F-13As were operated by XX Bomber Command's 1st Photographic Reconnaissance Squadron (PRS) at Hsinching, China, and by the 3rd PRS on Saipan. Later in the war they were based on Guam. The F-13As were used to gather pre and post strike photographs of targets and to assess landing beaches in Japan as part of the preparations for invasion should it have become necessary. In April 1944, a 3rd PRS F-13A appropriately named *Tokyo Rose* became the first US aircraft to fly over Tokyo since the Doolittle raid of 1942.

Postwar, the reconnaissance Superfortresses were redesignated initially as FB-29s and then RB-29s or RB-29As (depending on the version on which they were based) and also used in the Korean War.

WB-29: Weather reconnaissance conversion developed in 1946 for the USAAF's Air Weather Service, which provided global weather information for the Air Force. At least 80 aircraft operated in this role until 1958. The designation WB-29 was applied from 1950, RB-29s being used in this role before then. An Air Weather Service RB-29 flew over the North Pole in March 1947.

EB-29B: 'Mother ship' conversion to carry the McDonnell XF-85 Goblin parasite jet fighter on a retractable trapeze mechanism installed in the B-29's bomb bay and used to launch and recover the fighter. EB-29B 44-84111 was used in the trials, the first launch occurring in July 1948 and the first successful recovery the following October. Only two XF-85s were built and had the parasite fighter idea become operational, Convair B-36 bombers would have been used as mother ships.

ETB-29A: Another parasite fighter mother ship, ETB-29A 44-62093 was involved in extraordinary trials involving two Republic EF-84 Thunderjet fighters, one on each side of the bomber attached wingtip to wingtip. Trials proved the idea to be extremely dangerous and all three aircraft and their crews were lost in an April 1953 accident.

YB-29J: The conversion of six B-29s fitted with R-3350-CA2 fuel injected engines in modified nacelles. Substantial testing revealed some per-

F-13A Superfortress reconnaissance aircraft 42-24583, converted from a standard B-29.

Two shots of the EB-29B 'mother ship' (44-84111) with its McDonnell XF-85 Goblin parasite fighter in close attendance. The first launch occurred in July 1948 and the first successful recovery three months later.

formance gains but production was not undertaken. These aircraft were capable of very long ranges, one of them flying 7,916 miles (12,740km) non stop from Guam to Washington DC in November 1945. Two were subsequently converted to YKB-29J tankers and others were reconfigured as RB-29J reconnaissance aircraft.

XB-39: Conversion of the first service test YB-29 (41-36954) to take four 2,600hp (1,940kW) Allison V-3420 liquid cooled inline engines for trials by the manufacturer. This complicated powerplant was essentially two V-1710s (as fitted to the Curtiss P-40 and Lockheed P-38) driving a single propeller. The Allison engines gave the XB-39 the much improved maximum speed of 405mph (652km/h) at 35,000 feet but production did not go ahead.

XB-29E: Designation given to a single aircraft used to test a revised fire control system.

B-29F: Six B-29s modified to undertake cold weather trials in Alaska.

XB-29G: Conversion of Bell Atlanta built B-29B 44-84043 as a jet engine testbed with the engine mounted on retractable cradle in the bomb bay. This cradle could be extended in flight to offer the test engine to the airflow, and it could be started in flight. Engines tested on the XB-29G included the Allison J35, Pratt & Whitney J42 and J48, and General Electric J47 and J73. Another Superfortress engine testbed was B-29 45-21808 which was used to trial a ramjet suspended from the rear bomb bay.

XB-29H: One-off conversion of a B-29A for special armament trials. Several B-29s were used for this role,

The quite extraordinary ETB-29A Superfortress (44-62093) with its two EF-84 Thunderjet parasite fighters attached to the wingtips. All three aircraft were lost during a test flight.

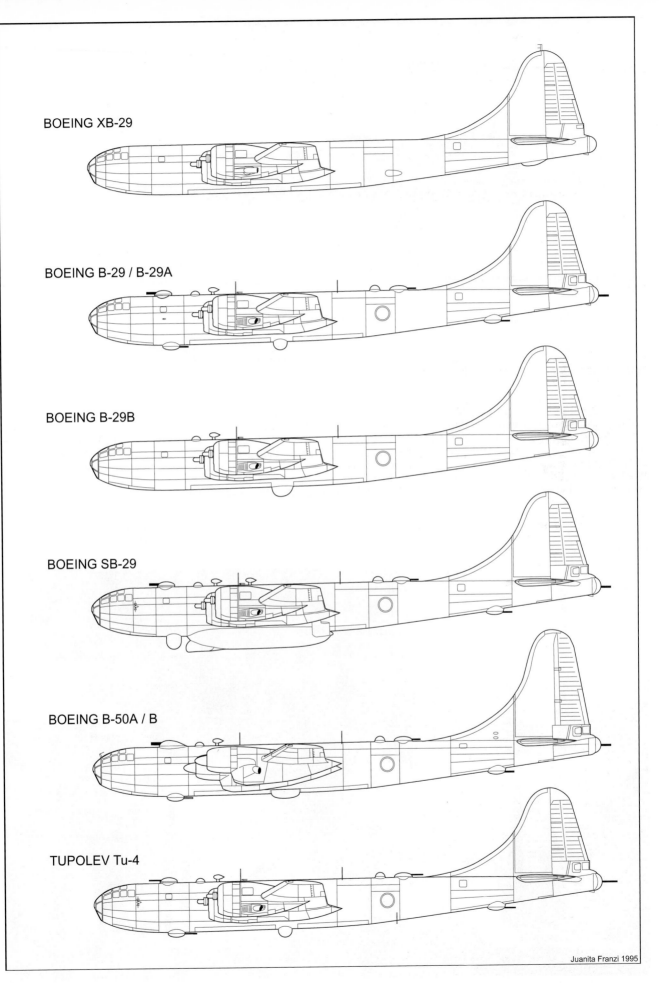

BOEING XB-29

BOEING B-29 / B-29A

BOEING B-29B

BOEING SB-29

BOEING B-50A / B

TUPOLEV Tu-4

Juanita Franzi 1995

The sole XB-29G engine testbed (44-84043) converted from a B-29B. The test engines were mounted on a retractable cradle fitted in the rear bomb bay. (Boeing)

including 42-24441 in late 1944 to test various manned turret configurations and another which carried two British 22,000lb (9,980kg) Grand Slam bombs under its wings. Another one carried a single Grand Slam partially internally under the fuselage.

B-29K: Designation originally allocated to aircraft converted to tankers but instead applied to one B-29 used as a cargo transport, subsequently as the CB-29K.

B-29L: Designation allocated to conversions capable of receiving fuel in flight but not used. B-29MR designation used instead (see below).

YKB-29J: Two conversions of YB-29Js (themselves engine test conversions of standard B-29s) to test the Boeing designed 'flying boom' aerial refuelling system in 1948. The system was subsequently adopted by the US Air Force as its standard method of transferring fuel and used in very large numbers on various tanker aircraft ever since.

YKB-29T: One off conversion (from B-29 45-21734) of a KB-29M (see below) with triple point hose and drogue aerial refuelling system.

KB-29M: Designation given to 92 B-29s converted to tanker aircraft from 1948. Aircraft were converted at the Boeing Wichita plant – which was reopened for the purpose – and used the first generation British trailing hoses and grapnel hooks system. The KB-29Ms served with the USAF's 43rd and 509th Air Refuelling Squadrons.

KB-29MR: The 'receiver' partner to the KB-29M (thus the 'R' in the designation); 74 conversions performed.

The one-off XB-29H was used for special armament trials while other aircraft used for similar roles retained their original designations. This underside shot of one of them (taken in August 1945) shows a British 22,000lb (9,980kg) Grand Slam bomb mounted partially internally under the fuselage. (Boeing)

A YKB-29J tanker for testing the flying boom system. The aircraft is hooked up to a Republic F-84 Thunderjet fighter. (Boeing)

KB-29P: The definitive B-29 tanker variant, fitted with the Boeing flying boom system. The first KB-29P was delivered in September 1950 and a total of 116 conversions was performed. The first operator of the KB-29P was the 97th Air Refuelling Squadron (ARS) at Biggs AFB, Texas.

Two KB-29Ps of the 27th ARS at Bergstrom, Texas, achieved some immortality on 25 November 1957 when they were retired as the last Superfortresses to serve the USAF in an operational capacity. The very last

USAF B-29 sortie of any kind – a radar evaluation flight – took place on 21 June 1960.

SB-29 'Super Dumbo': Air-sea rescue variant, the origins of which went back to the Pacific War when B-29s with liferafts and survival equipment were used to help crews which had ditched on the long bombing raids between the Marianas and Japan. The SB-29 carried a jettisonable A-3 liferaft and other survival equipment and remained in service until 1956.

DB-29: Drone director conversion.

GB-29: 'Mother ship' for some of the experimental high speed X-planes of the late 1940s/early 1950s. A conversion of B-29 45-21800, this aircraft achieved considerable fame in 1947 when it acted as the host for the Bell X-1, the first aircraft to exceed Mach 1.

QB-29: Pilotless target drone.

TB-29: Operational trainer variant converted in small numbers.

XB-44: Conversion of B-29A 42-93845 to take four 3,500hp (2,610kW) Pratt & Whitney R-4360 Wasp Major radial engines, and as such acted as the

Several KB-29 Superfortress tanker subvariants were created by conversion, the most numerous of which was the KB-29P fitted with the Boeing designed flying boom system. These aircraft belong to the 509th Air Refuelling Squadron. (Boeing)

An SB-29 'Super Dumbo' air-sea rescue aircraft with droppable A-3 lifeboat in place under the fuselage. (Boeing)

A close up shot of the Boeing flying boom installed in a KB-29. This system was adopted as standard by the US Air Force, while the US Navy went with the British hose and drogue method. (Boeing)

The GB-29 Superfortress 'mother ship' for the Bell X-1 and X-2 experimental high speed aircraft. This aircraft (45-21800) launched the X-1 when it became the first aircraft to exceed Mach 1 in 1947.

prototype for the B-50 Superfortress series. A successful test programme by the XB-44 led to the USAAF ordering a production version in July 1945, initially as the B-29D but subsequently as the B-50 (which see).

P2B-1: US Navy designation for four ex-USAF aircraft as test beds.

Washington B.1: The British applied the name 'Washington' to 88 B-29s loaned to the Royal Air Force in 1950 as a stopgap pending the arrival of sufficient quantities of English Electric Canberra jet bombers. The aircraft were allocated the RAF serials WF434-448, 490-514, 545-574; WW342-356; and WZ966-968.

They served with nine RAF squadrons between 1950 and 1958, although most had been returned to the USA by 1954. The last RAF squadron to fly the Washington – No 192 – used its aircraft on a variety of electronic warfare and electronic intelligence gathering (Elint) tasks.

Two RAF Washingtons found their way into Royal Australian Air Force (RAAF) service In 1952. These aircraft retained their RAF serials (WW353 and WW354) whilst in Australia, flying with the Aircraft Research and Development Unit Trials Flight in support of the British Ministry of Supply at the Woomera Rocket Range in South Australia. Both were sold for scrap in 1957.

The XB-44 (42-93845) conversion from a B-29A fitted with Pratt & Whitney R-4360 Wasp Major engines and in effect the prototype for the B-50 Superfortress series. (Boeing)

Four ex USAF B-29s flew in US Navy colours as the P2B-1. They were used as testbeds including as a 'mother ship' for the rocket and turbojet powered Douglas Skyrocket. (Boeing)

One of 88 B-29s delivered to the Royal Air Force in 1950 as a stopgap pending the arrival of Canberra jet bombers. In RAF service the aircraft was called the Washington B.1. (Boeing)

The Tupolev Tu-4, an unashamed and unlicensed Russian copy of the B-29. About 1,200 were built.

The Tupolev Tu-70 airliner development of the Tu-4 with cabin windows, a stepped windscreen design and accommodation for 72 passengers. (Boeing)

The Soviet Superfortress

The old USSR was never too modest to copy the West's technology and produce it without the benefit of a licence. The B-29 Superfortress was subject to such an 'arrangement' following the emergency diversion of three USAAF aircraft to an area near Vladivostok in 1944. These aircraft were engaged on operations over Japanese targets and the first of them landed on Soviet soil in July 1944 after running short of fuel.

The Soviet government interned the B-29s and handed them over to the Tupolev design bureau which immediately set about copying the aircraft under the designation Tu-4. The NATO codename *Bull* was later applied. The advanced technology in-

corporated in the B-29 gave the Soviet aircraft industry a wealth of useful information about systems, structures and radar and the unexpected arrival of the three aircraft was very timely as it coincided with Russia gaining access to the West's atomic bomb technology.

This and the subsequent 'Cold War' era was one in which everything Soviet was regarded by themselves as the best, anything from the West was rubbish and they invented anything worthwhile. This is well illustrated by a Russian book on the Tu-4 which fails to mention – even once – the Boeing B-29 from which the design was stolen!

Powered by four 2,000hp (1,490kW) Shvetsov Ash-73TK radial engines

(themselves based on the B-29's Wright R-3350s), some 1,200 Tu-4s were manufactured of which about 100 were supplied to China in 1951 to equip that country's newly established strategic bombing force. China also converted at least two Tu-4s to 4,000shp (2,985kW) Ivchenko AI-20 turboprops from an Antonov An-12 transport and adapted one to carry reconnaissance drones and another as the early warning AP-1 with large rotating radome.

Tupolev also developed a 72 passenger airliner development of the Tu-4, the Tu-70 *Cart*, with stepped windscreen and passenger windows, but this did not enter production.

The Tu-4 was a very important aircraft in the development of Soviet

China received about 100 Tu-4s from the Soviet Union of which a couple were converted to Ivchenko turboprop power. This example is preserved at the Museum of Chinese Aviation, Beijing. (Brian Candler)

strategic aviation as it served as the starting point for a long line of development culminating in the Tu-95 and Tu-142 *Bear* series of very large, swept wing turboprop powered bombers and reconnaissance aircraft as well as the Tu-95's airliner equivalent, the Tu-114.

B-50 SUPERFORTRESS

Built in relatively modest quantities with 370 examples emerging from Boeing's Renton plant between 1947 and 1953, the B-50 Superfortress resulted from investigations into an improved B-29 offering better payload capabilities and eliminating some of the deficiencies which persisted with the earlier aircraft. Although intended as a bomber, most B-50s were converted to other roles, notably reconnaissance and as tankers. As a bomber, the B-50's service was limited as its arrival coincided with the early days of jet bomber deployment in the USAF.

Work on the project began in 1944 under the Boeing designation Model 345-2, the fundamental areas looked at being increased power and operating weights plus weight savings to keep the empty weight of the bomber as low as possible.

The powerplant chosen for the Model 345-2 was the Pratt & Whitney R-4360 Wasp Major 28 cylinder, four row, air cooled radial engine rated at 3,500hp (2,610kW) with water injec-

tion. This engine therefore offered no less than 59 per cent more power than the B-29's Wright R-3350s. A B-29A (42-93845) was allocated to Pratt & Whitney for flight testing of the engine under the designation XB-44.

These tests proved successful, and that in combination with the other improvements developed by Boeing resulted in the USAAF ordering 200 aircraft in July 1945 as the B-29D.

Apart from the different powerplants, the B-29D featured several substantial changes over its predecessors including the use of more durable 75-S aluminium alloy instead of 24ST in the structure, redesigned engine nacelles to accommodate the new engine, strengthened undercarriage to cope with gross weights which at 168,700lb (76,522kg) initially was some 20 per cent greater than the B-29, and an increase in the height of the fin and rudder by five feet (1.5m). Major dimensions and layout were similar to the original B-29.

The B-29's internal bomb load capacity of 20,000lb (9,072kg) – either conventional or nuclear weapons – was retained, although provision was made for the carriage of an additional 8,000lb (3,630kg) on external racks under the inner wings. Defensive armament remained similar to the original B-29 with five turrets in the same positions as before and the computerised gunnery control system retained. Thirteen 0.50in machine guns

were housed in the turrets, four in the forward upper installation, three in the tail and two on each of the remaining turrets. The aircraft subsequently converted from bomber to other configurations lost some or all of this armament.

Political Manoeuvring

The end of hostilities with Japan in September 1945 saw the wholesale cancellation of most military aircraft orders including those which had been placed for the B-29. The B-29D order was reduced from 200 to 60 and in December 1945 the new bomber's designation was changed to B-50 by the Air Force. Ostensibly, this reflected the changes in the new version but it was really a piece of political sleight-of-hand intended to create the impression that this was a new aircraft rather than a development of an existing one which had just been subject to large scale order cancellations. In late 1945 the politicians were more inclined to allocate funding for new projects than upgraded old ones, and the redesignation of the B-29D to B-50 created the proper impression.

Interestingly, this situation was reversed just a few years later when the powers-that-be were often more generous with funding when improved versions of existing aircraft were proposed rather than entirely new ones. A good example is the F-86D Sabre

A KB-50J Superfortress tanker during a visit to Australia with a RAAF 'long nose' Lincoln Mk.31 on approach to land behind. (RAAF)

Some detail of the B-50's Pratt & Whitney Wasp Major engine installation in this shot, with the personnel on the ground giving scale to the aircraft. (Boeing)

B-50 Superfortress production was very modest by comparison with its predecessor, only 370 aircraft leaving the Renton production line between 1947 and 1953. (Boeing)

47-118, the first of 45 B-50B Superfortresses. (Boeing)

all weather fighter which was originally designated F-95 by the US Air Force as it was vastly different from the F-86 day fighter models which had preceded it. Political expediency dictated that the 'Sabre Dog' should also be an F-86!

The B-50 Superfortress was developed at a time when the USA's air forces were going through a period of reorganisation not just related to the end of World War II and the changes which inevitably followed. In March 1946 the USA's bomber forces came under the control of the newly established Strategic Air Command (SAC) and in September 1947 the United States Army Air Force (USAAF) finally achieved complete independence from the army to become simply the United States Air Force (USAF).

B-50A SUPERFORTRESS

There was no prototype B-50 as such, and the first production B-50A (46-002) was flown for the first time on 25 June 1947. Early aircraft were handed over for service testing in October 1947 and Strategic Air Command's (SAC) 43rd Bomb Group at Davis-Monthan AFB, Arizona, became the initial operator of the B-50A in February 1948.

One of the 43rd BG's B-50As (46-010 *Lucky Lady II*) undertook a momentous flight in March 1949 when it performed the first non stop aerial circumnavigation of the world, logging a distance of 23,452 miles (37,741km) in a time of 94hrs 1min, an average speed of 249mph (400km/h). The flight was important in that it proved that an SAC aircraft could reach any point in the world quickly and also established that the still young art of aerial refuelling was operationally viable. The B-50A was accompanied by KB-29M tankers of the 43rd Air Refuelling Squadron.

B-50A production reached 79 aircraft, the last of them delivered in January 1949. An additional aircraft (the 60th off the line) had been set aside for completion as a YB-50C with 4,500 (3,355kW) Pratt & Whitney R-4360 engines but the project was cancelled before the aircraft was built.

Eleven B-50As were converted to TB-50A crew trainers for the Convair B-36 Peacemaker bomber.

B-50B SUPERFORTRESS

The first of 45 B-50B Superfortresses (47-118) appeared in December 1948, differing from its predecessor mainly in having an increased maximum weight of 170,000lb (177,112kg). The B-50Bs had very brief lives as bombers, all but one being converted to RB-50B photographic and electronic reconnaissance aircraft for service with the 55th Strategic Reconnaissance Group.

All but one of the B-50Bs were converted to RB-50 photographic and electronic reconnaissance aircraft.

Of these 44 aircraft, 43 were subsequently converted again to other sub variants in 1950-51: 14 as RB-50E photo-reconnaissance aircraft; 14 as RB-50Fs with SHORAN navigation radar and 15 as RB-50Gs, similarly equipped as the 'F' but featuring the different nose cone shape of the B-50D.

The sole B-50B which escaped conversion to RB-50 configuration was the first example, 47-118, which was experimentally fitted with a caterpillar landing gear system and redesignated EB-50B. It was also used for other tests and trials.

B-50D SUPERFORTRESS

The most numerous of the B-50 versions with 222 built between May 1949 and December 1950, the B-50D differed from its predecessors in having a new one piece nose transparency, a receptacle (from the 16th aircraft) for the 'flying boom' air-to-air refuelling system which had been adopted by the USAF and a 700 US gallon (2,650 l) external tank under each outer wing allowing an unrefuelled range of 4,900 miles (7,885km). Gross weight grew again, this time to 173,000lb (78,473kg).

The B-50D also had a relatively brief career as a pure bomber as the Boeing B-47 Stratojet jet began to enter service in numbers. At their peak in 1951-52, about 220 B-50As and Ds were flying with five SAC Bomber Groups. All had swapped their by now obsolete B-50s for B-47s by the end of 1955.

The B-50D spawned several sub-variants by way of conversion, the most important of which were those rebuilt for aerial refuelling, not for use by Strategic Air Command but by Tactical Air Command (TAC), Pacific Air Forces (PACAF) and the US Air Forces in Europe (USAFE). SAC had access to its own fleet of tankers – Boeing KC-97s in the early and mid 1950s and Boeing KC-135 jets after that.

The first two B-50 tanker conversions were dubbed KB-50D and

The B-50D was the most numerous of the B-50 variants with 222 built in 1949 and 1950. About half were converted to KB-50J tankers.

A KB-50J tanker on the ground (top) and another example (bottom) with a trio of North American F-100 Super Sabres in attendance. Hayes Aircraft of Alabama converted 112 B-50Ds to KB-50J standards.

(below) The prototype XC-97 Stratofreighter with B-29 engines and tail surfaces. It first flew in November 1944.

these served as proof-of-concept aircraft for the major modification programme which was to follow. 'Production' Superfortress tankers were given the designation KB-50J, the first of which was flown in December 1957.

The KB-50J's most obvious external difference was the installation of a pair of General Electric J47-GE-23 auxiliary turbojet engines mounted in underwing pods outboard of the engines. Producing 5,620lb (25kN) thrust each, these engines served two purposes: two help the even heavier (179,500lb/81,420kg) KB-50J's take-off performance; and to give it a speed boost for more rapid deployment if necessary. A standard B-50D was capable of reaching a top speed of 380mph (611km/h) while the KB-50J could reach 444mph (714km/h) with all six engines at full power. The aircraft's cruising speed also received a 90mph (154km/h) boost thanks to the jets.

The KB-50J's refuelling equipment comprised three hose reels, one in the tail and two others in underwing pods near the tips, giving the ability to refuel three aircraft simultaneously. One hundred and twelve B-50Ds were converted to KB-50J configuration by Alabama based Hayes Aircraft (these days known as Pemco and well known for its freighter conversions of jet airliners), plus 16 TB-50Hs (see below) which as KB-50Ks were rebuilt to a similar standard. The last B-50 tanker was retired in the mid 1960s.

A front on view shows well the C-97's 'double bubble' fuselage, the lower lobe of which is from the B-29.

BOEING B-50 SUPERFORTRESS PRODUCTION SUMMARY

Notes: Total B-50 production amounted to 370 aircraft between 1947 and 1953 comprising 79 B-50As, 45 B-50Bs, 222 B-50Ds and 24 TB-50Hs. All were built at Boeing's Renton plant.

Model	USAF Serials	Qty
B-50A	46-002/060	59
B-50A	47-098/117	20
B-50B	47-118/162	45
B-50D	47-163/170	8
B-50D	48-046/127	82
B-50D	49-260/391	132
TB-50H	51-447/470	24

Other B-50D conversions included DB-50D and JB-50D special test aircraft, TB-50D crew trainers (11 conversions) and WB-50D weather reconnaissance aircraft with Doppler radar and temperature/humidity recording equipment (70 conversions).

TB-50H SUPERFORTRESS

The last new build version of the B-29/50 line, 24 TB-50Hs were manufactured, the last of them handed over in March 1953. Intended as unarmed bombing and navigation trainers, 16 were converted to three reel tanker configuration with auxiliary turbojets as the KB-50K.

A production C-97 Stratofreighter with B-50 tail and powerplants. The civil Stratocruiser was similar.

B-50D SUPERFORTRESS

Powerplants: Four Pratt & Whitney R-4360-35 Wasp Major 28 cylinder, four row radial engines with General Electric CH-7A exhaust driven turbo-supercharger each rated at 3,500hp (2,610kW) at takeoff with water injection and 2,650hp (1,975kW) normal; four bladed Curtiss Electric constant-speed, feathering and reversing propellers of 17ft 2in (5.23m) diameter.

Dimensions: Wing span 141ft 3in (43.05m); length 99ft 0in (30.17m); height 32ft 8in (9.96m); wing area 1,768sq ft (164.24m²)

Weights: Empty 81,000lb (36,742kg); max loaded 173,000lb (78,473kg).

Armament: Max bomb load 20,000lb (9,072kg) in two bomb bays; defensive armament 13 0.50in machine guns in five turrets.

Performance: Max speed 331kt (611km/h); cruising speed 241kt (445km/h); service ceiling 37,000ft (11,277m); max range 4,260nm (7,890km).

KB-50J supplementary data: Four 3,500hp (2,610kW) R-4360-35 piston engines plus two 5,620lb (25kN) thrust General Electric J47-GE-23 turbojets; length 105ft 1in (32.03m); empty weight 93,200lb (42,275kg); max loaded weight 179,500lb (81,421kg); max speed 386kt (714km/h) at 17,000ft; cruising speed 319kt (590km/h); initial climb (with jets) 3,260ft (993m)/min; ceiling 39,700ft (12,100m); range 2,000nm (3,700km).

One of two YC-97Gs, converted from standard aircraft by the fitting of 5,700ehp (4,250kW) Pratt & Whitney T34 turboprops.

The Transport Derivatives

To complete the B-29 and B-50 story, brief mention should also be made of the transport aircraft derived from the basic design, the military C-97 Stratofreighter and the Stratocruiser airliner.

Design work on a transport derivative of the basic design using the B-29's wings, tail surfaces, undercarriage and powerplants combined with a capacious new pressurised 'double bubble' two deck fuselage was undertaken in 1942 as the Model 367. The lower fuselage lobe was the same diameter as the B-29's. Three prototypes were ordered as the XC-97 Stratofreighter but because of Boeing's preoccupation with combat aircraft the first of these did not fly until November 1944.

Production C-97s differed from the prototypes in being based on the B-50 Superfortress, combining the new fuselage with the B-50's Pratt & Whitney engines, taller tail, increased weights and 75ST aluminium alloy structure. The first aircraft in the new configuration flew in March 1947 and total production of the C-97's many variants reached 888 before it ended in 1956.

Of these, 811 were built as KC-97 tankers with flying boom flight refuelling equipment, the aircraft becoming the USAF's standard tanker until the Boeing KC-135 jet came along.

Boeing also developed a civil airliner version of the aircraft as the Model 377 Stratocruiser. First flown in July 1947, only 56 were built and these served Pan Am, BOAC, SAS, American Overseas Airlines, United Air Lines and Northwest Airlines and others until replaced in most cases by more efficient propliners such as the Lockheed Super Constellation and Douglas DC-7.

Although capable of carrying up to 100 passengers, the Stratocruisers were most often used on the North Atlantic route between Britain and the USA with luxurious accommodation for a much smaller number of passengers complete with sleeper bunks and a bar/lounge on the lower deck, reached by way of a spiral staircase.

The Boeing Stratofreighter/Stratocruiser series lives on in the mid 1990s in the form of the surviving Aero Spacelines Guppy conversions of the 1960s and '70s with hugely enlarged upper fuselages for the carriage of outsized freight, initially spacecraft parts but more recently for airliner assemblies. Perhaps ironically, the best known operator of the Guppy has been Airbus Industrie, Boeing's main competitor on the world airline market. Airbus was still operating a small fleet of Guppies in 1995 for transporting assemblies for its airliners from factories throughout Europe to Toulouse for final assembly.

It could even be said that without the Guppy, Airbus would not have been able to establish the efficient manufacturing and assembly system it did, allowing it to challenge Boeing's position as the world's number one airliner manufacturer.

XB-29 Superfortress 41-002 first prototype at time of first flight on 21 September 1942.

B-29 Superfortress 'T N Teeny II' of 313th BW/9th BG USAAF, Tinian 1945.

B-29-BW Superfortress 42-24695 'Lucky Leven' of 498th BG USAAF, Saipan 1945.

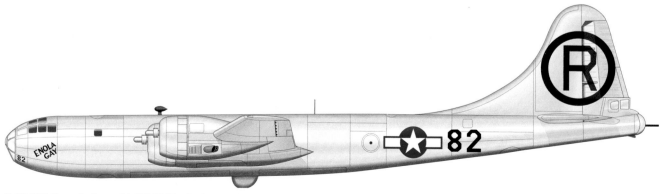

B-29-MO Superfortress 44-86292 'Enola Gay' of 509th Composite Group/393rd CS USAAF. Hiroshima atomic bomber flown by Col Paul Tibbets with false 6th BG tail markings.

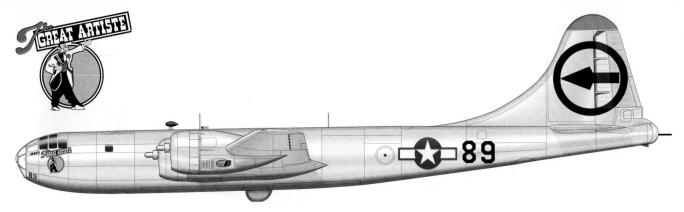

B-29-MO Superfortress 44-27354 'The Great Artiste' of 509th Composite Group USAAF. Accompanied Hiroshima atomic raid August 1945.

B-29 Superfortress 'Jostlin' Josie' of 498th BG/873rd BS USAAF, Saipan 1944. The first B-29 to land on Saipan.

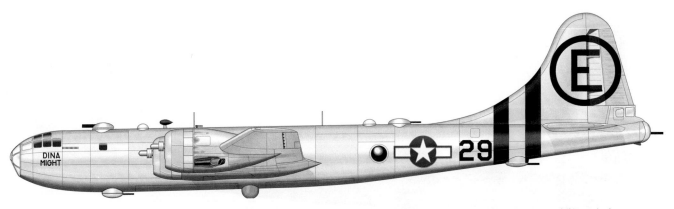

B-29-MO Superfortress 42-65280 'Dina Might' of 504th BG USAAF, Guam 1945. Original pinup artwork was removed 'by order'.

F-13A Superfortress 42-24621 'Yokohama Yo Yo' of 3rd Photo Reconnaissance Squadron USAAF, Saipan and Guam mid 1945.

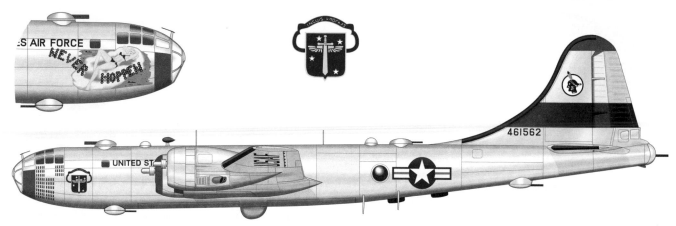

B-29A-BN Superfortress 44-61562 'Never Hoppen' of 19th BG/28BS USAF, Kadena Japan for Korean War operations. The artwork (originally nude on a bed) went through three stages of ever increasing modesty until it met with the CO's wife's approval!

B-29-BW Superfortress 45-21745 'Lucifer' of 19th BG USAF, Korea. Note Tarzon radio controlled bomb.

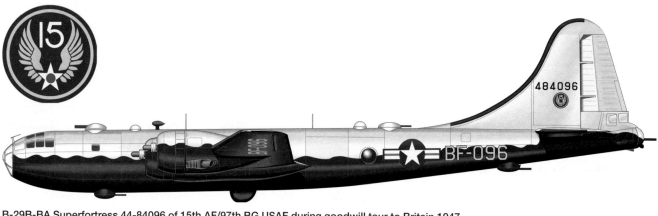

B-29B-BA Superfortress 44-84096 of 15th AF/97th BG USAF during goodwill tour to Britain 1947.

SB-29B-BA Superfortress 'Super Dumbo' 44-84078 of 3rd SRS USAF, Johnson Airbase Japan during Korean War, with droppable A-3 lifeboat.

WB-29A-BN Superfortress 44-61734 of 524th Weather Squadron USAF, Guam 1951. Squadron tracked weather from Guam over Japan to a point near Russia in support of Korean War operations.

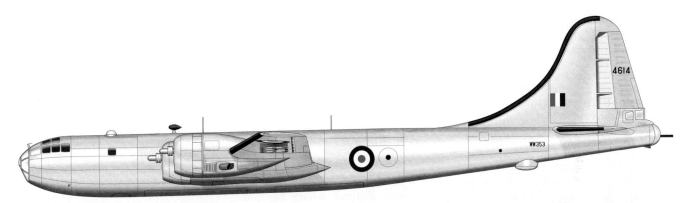

B-29 Washington B.1 WW353, RAF. This and another Washington transferred to RAAF in 1952 to fly in support of weapons trials at the Woomera Rocket Range, South Australia. Retained RAF markings in Australia.

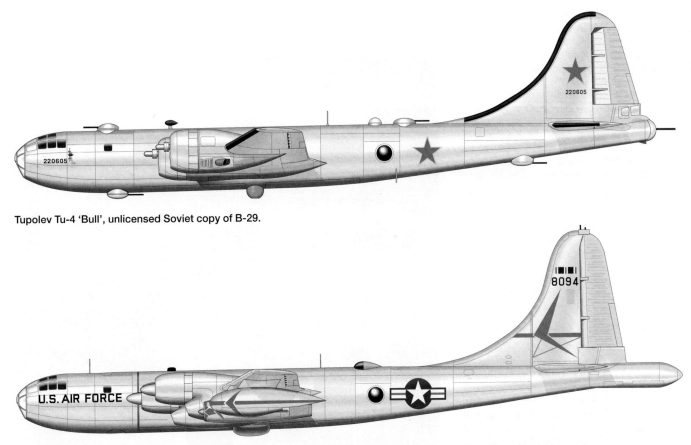

Tupolev Tu-4 'Bull', unlicensed Soviet copy of B-29.

KB-50J Superfortress 48-094 (converted from a B-50D) of 420th Flight Refuelling Squadron USAF, United Kingdom 1951.

SUPERFORTRESS AGAINST JAPAN

CHINA-BURMA-INDIA

The B-29's production, reliability and general organisational problems had only been partially overcome by the time the first aircraft departed for the new bomber's initial combat assignment, to the China-Burma-India (CBI) theatre for what had been called Project Matterhorn: using the B-29 to bomb Japanese steel plants and other industrial targets from bases in India and China.

The logistics of the operation were substantial. The basic plan was to operate the four Bombardment Groups of the 58th Bomber Wing from advanced bases around Chengtu (also spelt Chengdu) in south-central China, with main bases in India providing supplies and support. This necessitated the airlifting of fuel, ammunition and all other supplies over the 'Hump' (the Himalayas) from the Indian bases to China, in itself a major operation which inhibited the effectiveness of the B-29's early combat history. It was quickly discovered that the highly advanced nature of the B-29 created its own problems in that the aircraft was far from self sufficient and needed a massive support infrastructure to keep it going.

For this reason alone operations in the CBI were less than successful and it wasn't until later in 1944 when the B-29s were able to operate from the Marianas Islands that their full potential was realised, and even then after a shaky start.

The B-29 support operations in the CBI involved massive preparations. The new airfields around Chengtu were built at Kwanghan, Kuinglai, Hsinching and Pengshan as the B-29s' forward operating bases. Work on them had begun in November 1943 using local labour – a village quota of 50 workers per 100 households was imposed – and overall some 200,000 Chinese were used in the programme. Despite the available labour and the use of American military construction teams to supervise and organise, work proceeded at a slower pace than anticipated and it wasn't until May 1944 that they were just about usable.

The four bases in India – Chakulia, Charra (initially, then Dudhkundi), Piardoba and Kharagpur, the latter west of Calcutta, provided the headquarters for the B-29s' parent organisation, XX Bomber Command. The Indian airfields had been prepared in 1942-43 as bases for B-24 Liberators and although operational by the time the B-29s arrived were not fully prepared for them. An important element was the extension of the runways from 6,000 to 7,200 feet (1,830 to 2,195m) in order to allow the B-29s to operate at maximum weight. This was not done for a while, restricting the bombers' takeoff weights and thus their fuel and/or payloads.

Organisation

The CBI B-29s were under the overall control of XX Bomber Command with General Kenneth B Wolfe as Commander-in-Chief. The B-29 force was operated by the 58th Bombardment Wing and consisted of the 40th (based at Chakulia), 444th (Charra and Dudhkundi), 462nd (Piardoba) and 468th (Kharagpur) Bombardment Groups, each comprising four Bombardment Squadrons.

A basic reorganisation took place in late August 1944, less than three months after the first operations were flown, when Gen Wolfe was replaced by Major General Curtis E LeMay, fresh from Europe where he had made a major contribution to the tactics developed by the 8th Air Force in that theatre of war. He had risen quickly to the rank of Major General and was widely regarded as a superb air leader.

LeMay would have an immediate effect on the fortunes of the 58th BW's B-29s which had not been performing brilliantly under the command of Wolfe. LeMay's influence could not solve all the problems, but he made the 58th BW as efficient as it could be under the difficult conditions imposed by the China-Burma-India theatre.

One area which LeMay changed was the organisation of the squadrons. Each of the 58th BW's four Groups would now comprise three squadrons of ten aircraft rather than four with seven, in an attempt to simplify administration and control.

One other important organisational aspect which affected XX Bomber Command's operations was the high degree of autonomy which had been secured for it by Army Air Force chief General 'Hap' Arnold. The Command was at the spearhead of a newly established special strategic force, the 20th Air Force. Its brief was to undertake "the earliest possible progressive destruction and dislocation of

There's a healthy dose of symbolism in this photograph of Mount Fujiyama, taken from a raiding B-29. Until the advent of the Superfortress, Japan regarded itself as just about immune from attack by heavy bombers.

the Japanese military, industrial and economic systems and to undermine the morale of the Japanese people to the point where their capacity for war is decisively defeated". Arnold had exclusive use of the strategic bomber force and could assign it at his discretion.

This level of independence was unprecedented in the annals of American military aviation and helped set the scene for the establishment of the postwar US Air Force, no longer part of the Army. The cost was the need to demonstrate its capability, resulting in much pressure being put on Wolfe (while he still commanded the XXth BC) to get the B-29s into combat operations as quickly and effectively as possible. The circumstances and difficulties surrounding the CBI adventure did little to help.

A Problem of Logistics

The B-29s began flying to India from the USA in early April 1944 and by the first week of May some 130 had been deployed, the aircraft travelling to their bases via Marrakesh, Cairo and Karachi. Bombing operations were supposed to begin on 1 May, but this deadline proved impossible to meet due to the logistics in-

volved. The venture got off to a poor start when five B-29s crashed within a week of each other near Karachi during the early deployment flights.

Vast quantities of fuel, spare parts, bombs and other equipment had to be in place at the Chengtu airfields before any missions could be flown, and this all had to be moved 1,000 miles (1,610km) over the 'Hump' from the main bases in India. This in itself was a difficult and dangerous flight and each one counted as a combat mission, often signified by the painting of a camel on the aircraft. Even though the 58th BW had its own fleet of transports (including C-87s, converted B-24 Liberator bombers), by the beginning of May only 1,400 tonnes of supplies had been delivered to Chengtu, well short of the required amount.

Transport aircraft from Air Transport Command units in the CBI were diverted to help out – much to the annoyance of local commanders – but most of the supply sorties were actually flown by the 58th BW's own B-29s. To perform this role, selected aircraft were stripped of armament (except tail guns) and fitted with extra fuel tanks which carried a further 2,600 US gallons (9,840 l) over and

above the aircraft's standard capacity. But even this was basically inefficient. In good weather conditions the B-29s burned two gallons of fuel for every one gallon they delivered, and if diversions were necessary to avoid bad weather over the Himalayas or if headwinds were encountered, it could take as much as 12 gallons of fuel to deliver one.

One further problem which faced the B-29 crews once combat missions got underway was that Chengtu was too far west, requiring a large proportion of the flight to be conducted over Japanese held territory. Even then, only the southern Japanese home island of Kyushu could be reached.

Operations Begin

The first B-29 operational mission finally took place on 5 June 1944 when 98 aircraft were dispatched from India to bomb the Makasan railway yards at Bangkok. The mission involved a 2,000 mile (3,200km) round trip and was far from the dream start everyone was hoping for. Fourteen aircraft aborted before reaching the target due to engine failures, an overcast made radar bombing necessary but despite this only 18

Allied leaders at the January 1943 Casablanca Conference, the meeting at which 300 bombers were promised to China's Generalissimo Chaing Kai-shek to help in the fight against Japan. The type of bomber was not specified, but it soon became obvious the only one capable of performing the long range missions involved was the B-29. Photographed at the meeting are (standing, left to right): Lt Gen 'Hap' Arnold, Adm Ernest King, Gen George Marshall, Adm Sir Dudley Pound, Gen Sir Alan Brooke, Air Chief Marshal Sir Charles Portal; (seated) Franklin Roosevelt and Winston Churchill.

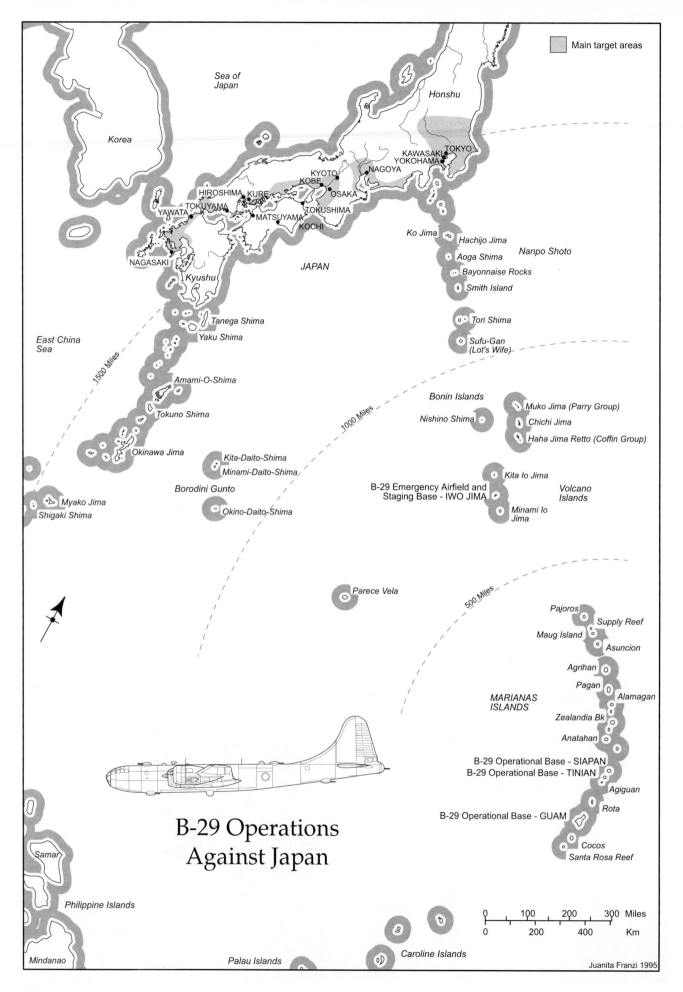

Main target areas

Sea of
Japan

Honshu

Korea

KAWASAKI TOKYO
YOKOHAMA
KYOTO NAGOYA
KOBE
HIROSHIMA KURE OSAKA
YAWATA TOKUYAMA TOKUSHIMA
MATSUYAMA
NAGASAKI KOCHI
Kyushu JAPAN

Ko Jima
Hachijo Jima
Aoga Shima Nanpo Shoto
Bayonnaise Rocks
Smith Island

East China
Sea

Tanega Shima Tori Shima
Yaku Shima Sufu-Gan
(Lot's Wife)

1500 Miles

Amami-O-Shima Bonin Islands
Muko Jima (Parry Group)
Tokuno Shima Nishino Shima Chichi Jima
Haha Jima Retto (Coffin Group)
1000 Miles

Okinawa Jima Kita-Daito-Shima Kita Io Jima
Minami-Daito-Shima Volcano
B-29 Emergency Airfield and Islands
Staging Base - IWO JIMA
Myako Jima Borodini Gunto Minami Io
Shigaki Shima Okino-Daito-Shima Jima

Parece Vela 500 Miles

Pajoros
Supply Reef
Maug Island
Asuncion

Agrihan

Pagan
Alamagan
MARIANAS
ISLANDS Zealandia Bk

Anatahan

B-29 Operational Base - SIAPAN
B-29 Operational Base - TINIAN

B-29 Operations
Against Japan

Agiguan

Rota

B-29 Operational Base - GUAM

Samar

Cocos
Santa Rosa Reef

Philippine Islands

| 0 | 100 | 200 | 300 Miles |
| 0 | 200 | 400 | Km |

Mindanao

Palau Islands Caroline Islands

Juanita Franzi 1995

bombs fell on the target area, 42 B-29s had to divert to various airfields as they ran short of fuel and five aircraft crashed on landing. The bombing runs were made from between 17,000 and 27,000 feet (5,200-8,200m) instead of the planned 22-25,000 feet (6,700-7,600m) due to the difficulties encountered in maintaining the specified four aircraft 'diamond' formations.

Nine days later, on the night of June 14/15, the first raid on Japan itself was launched when sufficient supplies were scraped together to allow 68 B-29s to attack the Imperial Iron and Steel Works at Yawata on the island of Kyushu. Once again, the results were very poor. Although 47 Superfortresses made it to the target this time, only one bomb fell within three-quarters of a mile of the aiming point. To make matters worse, six B-29s were lost in accidents and another was shot down by enemy fire.

Despite this, the raid attracted a great deal of publicity in the USA thanks to its propaganda value, although the sorry details were conveniently omitted!

This had an unfortunate side effect. No doubt stirred by the publicity generated by the Yawata raid, politicians in the USA wanted more and pressure was applied for this to happen. 'Hap' Arnold ordered Wolfe to send his B-29s "the length and breadth of the Japanese Empire". This was all very well, but there was a serious snag. The Yawata raid had left fuel stocks at Chengtu at less than 5,000 US gallons, not much more than half the capacity of a single B-29! Wolfe informed his superiors that it was impossible for their orders to be carried out and in early July he was recalled to Washington and reassigned.

The date 15 June was doubly significant in the B-29's history as it was also the day on which the campaign to capture the Marianas island of Taipan began. Within a few months, Taipan and nearby islands would be serving as bases for hundreds of B-29s.

The Arrival of LeMay

Wolfe's successor, Curtis LeMay, arrived eight weeks later, the 58th BW in the meantime carrying out several missions against industrial targets on Kyushu plus a steel manufacturing facility in Manchuria and oil storage tanks in Borneo, the latter operating from British bases in Ceylon. Some missions were flown at night, others in daylight and the apparently random nature of the targets selected suggested there was no central purpose, that the 58th BW's operations were being conducted in a desultory manner.

Curtis LeMay's arrival significantly changed that. A strong and purposeful leader with considerable experience and expertise developed in Europe, he immediately made changes which transformed B-29 operations in the CBI and which would help prepare their crews for the much more successful operations soon to be carried out from the Marianas.

LeMay decreed that all operations would be precision affairs conducted in daylight; that raids would be led by experienced crews who would be responsible for finding and marking the targets (along the lines of the RAF's Pathfinders); that both the bombardier and radar operator would control the bombing run so that whoever had the best sight of the target at the point of bomb release could perform that action; and that 12 aircraft box formations would replace the previous four aircraft diamond formations. All of these changes were the result of LeMay's European experience and although they made combat operations more effective, the fundamental flaw – the difficulty of supply – remained.

Results began to improve immediately with damage to the targets increasing. Losses, however, remained high as increased resistance was met from Japanese fighters and from light bombers which dropped phosphorous bombs on the B-29 formations. Japanese raids on the Chengtu bases added to the tally of B-29s de-

B-29 42-24579 'Eddie Allen' of the 40th BG, 45th BS in China during the Superfortress's first deployment. Note the camel symbols on the nose, indicating trips over the 'Hump' between India and China. Eddie Allen was the Boeing test pilot killed in the second prototype XB-29. (Boeing)

B-29s of the 468th BG unleash their payload on Rangoon in 1944.

stroyed, with the result that by mid December 1944 the significant total of 147 aircraft had been lost (80 in combat, the remainder from other causes), equivalent to the 58th BW's entire strength just seven months earlier.

It became clear that even with LeMay's inspired leadership, flying the B-29 from India and China was an ill-starred adventure. The decision was therefore taken in December 1944 to wind up Chengtu operations and deploy the B-29s to the newly captured Marianas Islands in the Pacific.

The final mission flown from Chengtu was recorded on 15 January 1945 (against targets in Formosa) after which the B-29 units withdrew to their bases in India. From there they flew some tactical support missions and the final 'proper' raid was against oil storage facilities at Singapore on 29 March. The 58th BW went to the Pacific in April 1945 as part of XXI Bomber Command.

The B-29s' record in the CBI belied its real capabilities. To put it bluntly, they had achieved very little if anything and had failed to "justify the lavish expenditure poured out on their behalf", according to the US official history. They had conducted 49 missions comprising 3,058 aircraft sorties, dropped 11,400 tonnes of bombs and lost nearly 150 aircraft. Only two sorties per aircraft per month were managed and just 800 tonnes of bombs were dropped on Japan itself.

Better was to come ... eventually.

THE MARIANAS

Even before XX Bomber Command's Superfortresses had deployed to India and China, plans were in place to capture the Marianas group of Islands in the central Pacific. Comprising three main islands (Saipan, Tinian and Guam) plus numerous smaller pieces of land, the Marianas would provide US forces with a springboard from which they could reach Japan. Situated about 1,500 miles (2,400km) south-east of mainland Japan, the Marianas were just within the B-29's operational radius.

Flying from the Marianas provided several advantages compared with the operations conducted from China: the aircraft would be relatively safe from enemy defences until they reached Japan itself; the bases would be virtually immune from counterattack as Japanese forces were gradually pushed northwards; once the necessary support infrastructure had been established it would be easier to maintain as there was a direct supply route from the USA; and all of mainland Japan could be reached by the B-29s.

There were also several disadvantages, most of them stemming from the very long overwater flights involved. A badly damaged B-29 or one which found itself in a fuel emergency situation would inevitably end up ditching somewhere in the Pacific Ocean, and of course no fighter protection could be provided. The situation would later improve with the capture of Iwo Jima, midway between the Marianas and Japan. This gave damaged B-29s a 'halfway house' on which they could land if it was impossible to return to the Marianas and also provided a base from which escorting fighters could operate.

The Marianas Springboard

The campaign to take the Marianas began on Saipan on 15 June 1944 following a four day naval and air bombardment. The assault landing by US Marines was heavily opposed and cost 3,000 American lives. The Japanese lost 24,000 and by 9 June Saipan was completely in US hands. Guam and Tinian were captured within a fortnight. The size of the operation is put into some perspective when it is noted that the enterprise was carried out by a task force which was larger than the entire Japanese Navy! Such was America's military and industrial might.

A B-29 of the 314th BW/19th BG on Guam undergoes some open air maintenance during 1945. Note the Wright R-3350 engines stockpiled in the foreground.

Construction work on airfields and support facilities began on Saipan before the end of June, even before the island had been properly secured. It was realised that time was of the essence and that it was necessary to have a transition from India/China B-29 operations which was as 'seamless' as possible.

Regardless of this, the sheer volume of work involved meant that when the first B-29s began arriving at Saipan in October 1944 work was far from finished with the planned twin 9,000ft (2,740m) runways at Isley Field (based around a short coral airstrip built by the Japanese) existing only as a single 6,000ft (1,830m) sealed strip. There were also no proper hardstandings and buildings yet, despite an enormous effort by the US Navy engineers, the Seabees.

This necessary infrastructure would soon be up and running, however, with operational airfields situated on all three of the major islands, not just for B-29s but for other American military aircraft as well.

The B-29s which would be based in the Marianas were operated by XXI Bomber Command, established specifically for the purpose. It was commanded by Major-General Haywood Hansell Jr and in turn was controlled by the 20th Air Force commanded by Hap Arnold. The first Bomber Wing to be based in the Marianas (on Saipan) was the 73rd which consisted of four Bomb Groups (the 497th, 498th, 499th and 500th) each of which had three squadrons of 10 aircraft. Other Bomb Wings would subsequently join the fray and by the cessation of hostilities in August 1945 five of them were operating from Saipan, Tinian and Guam, represent-

The first B-29 base in the Marianas was established on Saipan, but the airfield and its associated facilities were far from ready when the first aircraft arrived in October 1944.

ing a strength of about 900 B-29s at a given time. The other Wings involved were the 313th (from January 1945), 58th (April), 314th (April) and 315th (June).

Five airfields were built to support the B-29s: Isley Field on Saipan; West Field, Tinian; North Field, Tinian; North Field, Guam; and Northwest Field, Guam. Tinian's North Field (opened in early 1945) was at the time the biggest bomber base ever constructed with four sealed parallel runways each with a length of 8,500 feet (2,590m).

Isley Field was named after US Navy Commander Robert H Isely (note the spelling) but his name had unfortunately been subject to a typographical error by an anonymous clerk somewhere along the line and the incorrect version remained in use.

Operations and Problems

B-29s of the 73rd BW began arriving on Saipan on 12 October 1944, the lead aircraft flown by XXI Bomber Command's CO, Haywood Hansell. The 73rd BW would have operations to itself for the first three months and it was decided by Hansell that these should comprise high altitude daylight precision bombing, with the emphasis switching from steel works and the like to targets which would produce a more immediate result, such as those directly contributing to the Japanese aircraft manufacturing programme.

The first shakedown raids were conducted from 27 October against tactical targets on the Japanese held island of Truk, part of the Caroline Islands group to the south of the Marianas. The results were all too familiar – inaccurate bombing, ragged formation keeping and an unacceptably high level of aborts, a worrying number of them due to a new spate of engine failures. A large part of the problem was the crews' training. They had been primarily trained to operate in a radar bombing environment, in which there was greater latitude of bombing heights and other factors. Also, training in this method of bombing implied night operations. High altitude precision bombing demanded a high level of skill which the crews had not as yet fully developed.

Several other raids on Truk in late October and into November produced similar results, and as if to rub further salt into the wounds, the Japanese launched several strikes against Saipan from Iwo Jima, damaging several aircraft. Retaliatory raids were quickly ordered.

There was pressure to begin operations against Japan but Hansell was reluctant to quickly indulge as

The writing of messages of greeting (?) on bombs before a mission is a tradition as old as the art of bombing itself. The words 'Tokyo Busters' and 'You'll Get a Bang Out Of This One' are obvious here.

USAAF heavies in the Pacific in July 1945 (left to right): Generals Twining, LeMay, Spaatz and Giles.

the base was still incomplete by mid November, with maintenance facilities in particular far from ready.

The first strategic raid flown by XXI BC's Superfortresses was on 24 November 1944 against the Nakajima Aircraft Company's Musashi engine factory just outside Tokyo. This factory was responsible for the production of about 30 per cent of Japan's aero engines, so its importance was great. One hundred and eleven B-29s set out to attack, but things immediately began to go wrong when no fewer than 17 aborted with engine problems. Even Mother Nature was against the American flyers. Approaching Japan at between 27,000 and 32,000 feet (8,200-9,700m) they hit the winter jetstream which came out of the west at that time of the year. The extremely strong winds blew the B-29s all over the place, making accurate bombing difficult if not impossible. The target was also covered in cloud and the result of all this was that only 24 B-29s dropped their bombs in approximately the right place, the remainder spreading

their's over various parts of Tokyo. One B-29 was lost, rammed by a Japanese fighter.

As had been the case with the first raid on Japanese soil the previous June by B-29s operating from China, this raid attracted a great deal of publicity at home, as it was the first on Tokyo since Doolittle's effort two years previous. Officially, the "Targets were successfully bombed as planned"!

Ten further raids were carried out on the Musashi plant over the next few weeks, employing the same tactics. These were no more successful than the first, the statistics speaking for themselves: only two per cent of the bombs hit the plant's buildings and only 10 per cent fell within the general area it occupied. The 73rd BW lost 40 B-29s, the 440 killed and missing crewmembers more than twice the number of Japanese casualties on the ground.

To rub it in, the US Navy's Task Force 58 attacked the Musashi plant in February 1945 using carrier borne fighter-bombers and light attack air-

craft. This single raid did more damage to the factory than all the B-29 efforts combined!

This was not how the plot had been written. The B-29 was supposed to be the ultimate symbol of air power, high technology and destruction, of the world's mightiest industrial nation. But so far, there had been little if any evidence it could take out an undefended village in the most underdeveloped country on earth, let alone bring the Japanese Empire to its knees. The political implications were also serious, as an independent US Air Force was hoped to be the reward for success.

The 73rd BW's second industrial target was the Mitsubishi engine factory at Nagoya which was raided several times in December. These efforts achieved slightly better results (17 per cent of the complex was destroyed) but at a far greater cost as more effective defences were claiming an average five B-29s per mission and the abort rate was almost one in four.

Steps were taken to improve the latter statistic, caused mainly by the extremely heavy weights at which the B-29s were operating. This in turn was straining their sometimes fragile engines. A weight loss programme was introduced involving removing one of the bomb bay fuel tanks and reducing the amount of ammunition carried. This reduced the empty weight of the aircraft by 6,000lb (2,720kg), resulting in a performance improvement. With the strain on the engines eased a little and with the introduction of better maintenance procedures, the time between overhauls of the R-3350s was increased from 250 to 750 hours and the failure rate declined. By mid 1945 the mission abort rate due to mechanical problems had been whittled down to about seven per cent.

A total of 414 B-29s was lost to enemy action and a further 104 to non combat causes during World War II. The photographs show a ditched 6th BG aircraft and another which has suffered a crash landing and subsequent fire.

LeMay's Second Coming

Despite these logistical improvements, the combat situation remained poor with no sign of improvement evident. Something needed to be done and it was – Hap Arnold decided to replace Haywood Hansell as Commander-in-Chief of XXI Bomber Command with the ubiquitous Curtis LeMay.

LeMay arrived in the Marianas in late January 1945, coincidentally at the same time as the second B-29 Bomb Wing, the 313th based at North Field, Tinian. This Wing made its operational debut in early February against Kobe in one of the last daylight precision bombing raids to be performed for a while.

LeMay had already studied the situation and faced with the facts, concluded that continuing precision attacks on specific industrial targets was not the way to go as the B-29 crews were generally not up to the task. Instead, he reasoned, advantage should be taken of the fact that Japan's cities were constructed mainly of timber buildings, buildings which burnt easily. The decision was therefore made to switch to a general urban bombing offensive which, it was reckoned, would be five times more effective than the previous tactics. For the moment, these raids would still be carried out at high altitude and in daylight.

The idea was first tried out in late January and early February against Nagoya and Kobe using 4,900lb (2,220kg) incendiary clusters with encouraging effect. In late February and early March Tokyo was attacked and the results were devastating, the first raid alone gutting 28,000 buildings after 172 B-29s dropped over 400 tonnes of incendiaries on the city. The pattern was already becoming established and within three months three-quarters of the B-29s' bomb loads comprised incendiaries. Particularly effective was the 500lb (227kg) 'pyrotechnic gel' bomb which had a mixture of jellied oil, petrol, magnesium powder, sodium nitrate and heavy oil. Fires started by the weapons were virtually impossible to extinguish and usually didn't go out until there was nothing left to burn.

The Offensive of Fire

March 1945 saw further revisions to the B-29's tactics and a boost to their operations generally with the taking of Iwo Jima after a bloody, four week battle. An airstrip was quickly built and fighter forces to escort the B-29s to Japan established. Iwo Jima also provided a mid point safe haven between Japan and the Marianas for damaged B-29s which could divert their if necessary. More

A B-29 of the 313th BW/505th BG drops its bombs over Japan. This Wing was responsible for most of the very successful mining operations around the Japanese coast and outside its major harbours.

than 2,400 Superfortresses did just that, saving many of the 25,000 or so crewmen aboard the aircraft who otherwise would have found themselves ditching in the Pacific.

The results of the early incendiary raids against Japan were carefully analysed by LeMay and his staff and several important revisions were made to the tactics employed as a result. The jetstream continued to be a problem so it was decided to switch to low altitude (between 5,000 and 9,000 feet) raids, many of which would be conducted at night. Many aircraft were also stripped of their heavy defensive gun system leaving only the tail installation intact. This further reduced the B-29s' operating weights, improving performance and reliability and increasing the aircraft's practical maximum bomb load from 12,000 to 18,000lb (5,440 to 8,160kg). The crews were at first suspicious of all this as they considered they were now more vulnerable to attack from Japanese fighters, but LeMay reasoned that the defences were geared to intercept incoming bombers at high altitude, there would be an element of surprise and that the cover of darkness would provide its own defence.

A new Superfortress variant which lacked all guns but the tail installation from the start was put into production in early 1945, the B-29B. Only 311 were built by Bell Atlanta and most of these served with the 315th BW which operated from Guam from June 1945. The 315th's main targets were petroleum, oil and lubricant facilities and most missions were flown at night.

The first raid utilising the new techniques was flown against Tokyo on the night of March 9/10. Led by pathfinder crews, 325 B-29s were launched of which 279 reached the target. The result was devastating for the city, which had 16 square miles (41sq km) destroyed in the accompanying firestorm, fanned by a 30mph (48km) winds. The philosophy of the 'Offensive of Fire' had arrived and at the first attempt had decimated the heart of Japan's capital city and killed 84,000 people. Twenty-five per cent of Tokyo's buildings were destroyed and one million were made homeless. This raid is regarded by many as the most destructive of the war, even more so than the atomic bomb missions. The cost to the USAAF was 14 B-29s, higher than hoped for but considered to be within reasonable limits

A similar raid on Nagoya two days later yielded similar results and further attacks on Osaka, Kobe and Nagoya once more over a two week period resulted in 120,000 killed and vast areas of the four cities reduced to rubble. At last the B-29 was starting to show some return on the investment, although it still had not been successful in the role for which it had been designed – high altitude precision bombing.

This pattern continued over the next few months with ever increasing numbers of B-29s participating as additional Wings joined the fray. Three hundred, 400 or even 500 bombers were sent out against single cities and for the first time in the history of strategic bombing there were signs of civilian panic on the ground.

The effect of the B-29s' incendiary raids is graphically illustrated by this shot of Tokyo after one such attack. Half of Tokyo's city area was destroyed by the 'fire raids'.

In Nagoya, some 170,000 people fled the city for the surrounding countryside in May after the B-29s had made yet another raid on them.

By the end of May all the major Japanese cities had suffered by the B-29s' fire raids. Tokyo, for example, had seen half of its city area destroyed. The 23rd of that month witnessed the largest B-29 raid of the war on a single target when no fewer than 562 aircraft headed for Tokyo, of which 510 reached their destination. Two days later another 464 bombers returned to attack areas which still had not been hit.

Tactical Variations

There were still concerns about losses. The two big Tokyo raids had cost 43 Superfortresses, so LeMay temporarily switched to high altitude attacks which would draw the Japanese fighters into combat with escorting American P-51 Mustangs from Iwo Jima. The idea was to drain their remaining strength and was largely successful as the Japanese decided to keep back most of what remained as a last ditch reserve for the seemingly inevitable invasion.

From early June 1945 the Americans had a virtual monopoly of the skies over and around Japan, grim proof being provided by a raid on Kobe on 5 June which was so effective the city was removed from the bombing list because it was considered there was nothing left worth attacking.

This complete freedom of the skies allowed LeMay to use his B-29s on more high altitude precision attacks against industrial targets using experienced crews. Experiments had begun as early as April when attacks on the old favourites – the aero engine plants at Musashi near Tokyo and Nagoya – proved to be successful. High explosives were used in both cases and the result was that the Japanese aero engine industry more or less ceased to exist.

Other raids against industrial targets were conducted between June and August 1945, resulting in the destruction of Japan's oil stocks and oil production facilities. The incendiary raids continued in the meantime against 58 smaller cities with populations between 100,000 and 200,000 people. Damage was considerable and the psychological war was also fought by B-29s which began dropping leaflets as well as bombs over the country.

With no fighter opposition of consequence to deal with, the US commanders were confident enough to give details of forthcoming raids to the Japanese people and sure enough,

B-29s of the 73rd BW/500th BG on an incendiary raid over Yokohama in May 1945. A total of 459 bombers took part in this raid, escorted by P-51 Mustangs from Iwo Jima.

the specified areas were devastated a few nights later. All this placed an enormous strain on the population and its government, both of which were starting to show signs of panic. By July the Imperial Cabinet was beginning to think about a negotiated end to hostilities, although the military hierarchy which had put Japan on the path to war in the first place remained defiant.

Another very important aspect of B-29 operations was the mining of Japanese home waters between late March and August 1945. These missions were flown mainly by aircraft from the 313th BW which during the campaign placed some 13,000 acoustic and magnetic mines in the western approaches to the Shimonoseki Strait and the Inland Sea as well as around the major harbours of Tokyo, Kure, Hiroshima, Nagoya, Aki, Noda and Tokuyama.

This proved to be a very effective tactic with virtually all Japanese coastal shipping coming to a standstill as a result. A breakout was ordered in May and the Japanese losses were heavy – 89 ships totalling 213,000 tonnes were sunk. Overall Japanese shipping losses due to the B-29s' mining activities were estimated at 800,000 tonnes.

By July 1945 there was little remaining in Japan for the B-29s to bomb. Industry had for all intents and purposes ceased to exist and vast urban areas of the cities had been destroyed. The next step inevitably appeared to be invasion, but the most profound event in the history of warfare would make that unnecessary.

The Atomic Bomb

That profound event was the dropping of atomic bombs on Hiroshima on 6 August 1945 and on Nagasaki three days later. Both bombs were dropped by B-29s operated by the 509th Composite Group flying from North Field, Tinian. The Superfortress has thus passed into history as the only aircraft to drop an atomic bomb in anger in the only use of the weapon under those circumstances.

By mid 1945 the US Government was faced with a dilemma as to what to do with Japan. It was obvious to everyone – including Japan's own leaders – that Japan could be destroyed by starvation and bombing without an invasion, but the hard core military leaders favoured fighting to the bitter and were successful in their campaign to prevent surrender.

The United States was therefore faced with two choices – either to invade with the risk of extending the war into 1946 at least and sustaining as many as one million allied casualties, or to try out a new weapon, one which promised to finally force Japan to sue for peace without the need for invasion. This was the atomic bomb, the most powerful weapon so far developed.

US President Franklin Roosevelt had died in April 1945, leaving his successor, Harry Truman with the decision as to which way to go. There was one other issue which Truman had to deal with, and that was concern about the Soviet Union and its intentions in the East. For the USA, it was desirable that the war against Japan be over before the Soviet Union had a chance to enter it. Stalin had not as yet declared war on Japan – and didn't do so until two days after the first atomic bomb was dropped on Hiroshima – but this action was considered inevitable now the battle had been all but won, and claims for territory and an involvement in the invasion and subsequent occupation were sure to follow.

At this stage the Soviet Union knew nothing about the Manhattan Project, the development of the atomic bomb. The project had been going ahead at full speed for three years by mid 1945 with two types of bomb being developed, one relying for its chain reaction on uranium ('Little Boy') and the other on plutonium ('Fat Man'). 'Little Boy' was 10 feet (3m) long, had a diameter of 28 inches (71cm) and weighed about 9,000lb (4,080kg) while 'Fat Man' was eight inches (20cm) longer, weighed 10,000lb (4,540kg) and was much larger in girth with a diameter of 60 inches (1.52m).

As early as mid 1943 Hap Arnold had been asked to supply modified B-29s so that flight and dropping

The 'Little Boy' atomic bomb was dropped on Hiroshima on 6 August 1945. The device weighed 9,000lb (4,080kg) and was 10 feet (3m) long. It produced an explosion equivalent to 20,000 kilotons of TNT.

tests of the new bombs (in dummy form) could be carried out. The modification of an early production B-29 began in December 1943 and the first drop test was performed at Muroc, California in late February 1944. Further modification of the bomb's installation and dropping mechanism was required and testing resumed in June 1944. Contracts for the modification of more B-29s were in the

meantime let and by August, 46 B-29s had been turned into 'atomic bombers'.

A special unit was established to operate these special B-29s, the 509th Composite Group based at Wendover, Utah. The 509th was unique in that it comprised but one squadron – the 393rd – and was a completely self contained and highly secret unit.

Its commanding officer was Col Paul Tibbets, an experienced pilot with combat over North Africa and Europe under his belt flying B-17s. He had most recently been working as a B-29 test pilot. The 509th began intensive training and by the northern hemisphere spring of 1945 was ready for overseas deployment. When it deployed to its Pacific base at North Field, Tinian in July 1945, the 509th was technically part of XXI Bomber Command, although it received its orders directly from the 20th Air Force. Upon arrival in the Marianas, further training flights were undertaken. Very few of the Group's personnel had any idea what the weapon they had been practising with actually was. Even if they did know they probably wouldn't have believed that such a thing was possible!

By then, no nuclear device had as yet been exploded but in February and March 1945 US commanders in the Pacific were told of the atomic weapon and its potential. On 16 July the first atomic device was exploded at Alamogordo, New Mexico, with spectacular results. Harry Truman quickly decided that in view of the potential cost of invasion and the Japanese military's reluctance to unconditionally surrender, the atomic bomb would be used. On 24 July General Spaatz, fresh from victory in Europe and commander of the newly established US Strategic Air Forces Pacific, ordered the 509th Composite Group to "deliver its first special bomb as soon as weather will permit visual bombing after 3 August 1945

Col Paul Tibbets in front of B-29 44-86292 'Enola Gay', the Hiroshima atomic bomber. (Boeing)

on one of the targets: Hiroshima, Kokura, Niigati and Nagasaki".

Components for 'Little Boy' began to arrive at Tinian on 29 July and by 2 August it had been assembled. On the same day, Curtis LeMay issued an order specifying Hiroshima as the primary target and Nagasaki and Kokura as alternates if the weather prevented a visual drop.

Armageddon

Two reconnaissance F-13 Super-fortresses departed Tinian in the early hours of the morning of 6 August to report on the weather conditions over the primary and secondary targets. Paul Tibbets' Martin built B-29 44-86292 *Enola Gay* (named after the commander's mother) and accompanying aircraft followed shortly afterwards with 'Little Boy' on board. The bomb was armed in the air after the target had been cleared by the preceding F-13As. At 17 seconds past 0815 hours, 'Little Boy' left the B-29's bomb bay. Tibbets initiated a 150 degree turn to put as much distance as possible between his aircraft and the explosion. Like everyone else involved, Tibbets had no idea what to expect from his special bomb, except that its explosion would be accompanied by a substantial blast and a high intensity flash.

The bomb was set to explode in the air at a height of 1,850 feet (565m) above the ground. It did, and 78,000 Japanese died instantly. Harry Truman said it was "the greatest thing in history".

The Hiroshima blast did not force Japan to surrender immediately, partly because of communications problems with Hiroshima but largely because its leaders could not comprehend what had happened. Also, the military hard core remained firm.

No official reaction had been received two days after the raid, so it was decided that the plutonium based 'Fat Man' – the only remaining atomic bomb in existence – should also be dropped.

The raid took place on 9 August with Kokura listed as the primary target and Nagasaki as the alternate. This time the B-29 carrying the bomb would be commanded by Captain Frederick Bock but flown by Major Sweeney. Appropriately, the aircraft was named *Bock's Car*. It was another Martin built B-29, serial number 44-27297.

The second atomic mission did not go off quite as smoothly as the first as Kokura was partially covered in cloud and the aiming point could not be seen after three attempts. The decision was therefore made to go to Nagasaki, which but for a fluke of nature would not have known what it

The atomic bomb's mushroom cloud billowed to 20,000 feet (6,100m) above Hiroshima. The photograph was taken from one of the B-29s accompanying 'Enola Gay' on the mission.

(below) Nose detail of the B-29 (44-27297) which dropped the plutonium based 'Fat Man' bomb on Nagasaki on 9 August 1945. The aircraft was named after its commander, Captain Frederick Bock.

feels like to be on the wrong end of an atomic bomb. Nagasaki was also largely covered by cloud but a break in it revealed the aiming point long enough for the bomb to be released at 1058 hours from a height of 28,900 feet (8,800m). The immediate death toll was about 35,000.

This time, Japan had no choice but to accept unconditional surrender, prompted by sensible intervention by the Emperor. Hostilities ended on 14 August and the formal surrender was signed aboard the battleship USS *Missouri* in Tokyo Bay on 2 September 1945. US General Douglas MacArthur received the surrender, ending the occasion with the words: "These proceedings are closed". Six years and one day after the German invasion of Poland had provoked the start of history's only truly global conflict, they most certainly and emphatically were.

The Nagasaki atomic raid was not the last mission performed by XXI Bomber Command's B-29s. Others using conventional weapons were carried over the following few days, culminating in a huge effort on 14 August, the final day of fighting. This raid had the air of a 'final fling' about it and involved no fewer than 804 aircraft attacking targets in various parts of Japan. It probably wasn't entirely necessary but did further ram the point home to Japan's leaders.

The vital statistics surrounding XXI Bomber Command's campaign against Japan include a total of 34,000 sorties flown, 160,000 tonnes of conventional ordnance dropped (of which 80 per cent fell between March and August 1945), 371 B-29s lost (all but 37 as a result of enemy action) and more than 3,000 crewmen killed, wounded

---

XXI BOMBER COMMAND B-29 UNITS PACIFIC 1944-45

58th Bomb Wing (West Field, Tinian from April 1945)
40th Bomb Group (25, 44 and 45 Bomb Squadrons)
444th Bomb Group (676, 677 and 678 Bomb Squadrons)
462nd Bomb Group (768, 769 and 770 Bomb Squadrons)
468th Bomb Group (792, 793 and 794 Bomb Squadrons)

73rd Bomb Wing (Isley Field, Saipan from October 1944)
497th Bomb Group (869, 870 and 871 Bomb Squadrons)
498th Bomb Group (873, 874 and 875 Bomb Squadrons)
499th Bomb Group (877, 878 and 879 Bomb Squadrons)
500th Bomb Group (881, 882 and 883 Bomb Squadrons)

313th Bomb Wing (North Field, Tinian from January 1945)
6th Bomb Group (24, 39 and 40 Bomb Squadrons)
9th Bomb Group (1, 5 and 99 Bomb Squadrons)
504th Bomb Group (398, 421 and 680 Bomb Squadrons)
505th Bomb Group (482, 483 and 484 Bomb Squadrons)

314th Bomb Wing (North Field, Guam from April 1945)
19th Bomb Group (28, 30 and 93 Bomb Squadrons)
29th Bomb Group (6, 43 and 52 Bomb Squadrons)
39th Bomb Group (60, 61 and 62 Bomb Squadrons)
330th Bomb Group (457, 458 and 459 Bomb Squadrons)

315th Bomb Wing (Northwest Field, Guam from June 1945)
16th Bomb Group (15, 16 and 17 Bomb Squadrons)
331st Bomb Group (335, 336 and 337 Bomb Squadrons)
501st Bomb Group (21, 41 and 485 Bomb Squadrons)
502nd Bomb Group (402, 411 and 430 Bomb Squadrons)

316th Bomb Wing (Kadena, Okinawa from July 1945)
333rd Bomb Group (435, 460 and 507 Bomb Squadrons)
346th Bomb Group (461, 462 and 463 Bomb Squadrons)

509th Composite Group (North Field, Tinian from July 1945, 393rd BS)

---

or missing. Japan's losses were enormous: more than 300,000 dead and 60 cities devastated with over 150 square miles (388.5sq km) of real estate laid waste.

SUPERFORTRESS OVER KOREA

The end of World War II brought with it massive cancellations of military equipment. The Superfortress did not escape, contracts for more than 5,000 of them being withdrawn in September 1945. Production continued for a time nevertheless, the last example emerging from Boeing's Renton factory in March 1946.

The US Army Air Force also underwent considerable organisational change. The bombers were reorgan-

ised into the newly established Strategic Air Command (SAC) in March 1946 while the USAAF itself ceased to exist in June 1947 with the establishment of the completely independent United States Air Force (USAF). The dream of years gone by had at last become reality.

The B-29 remained SAC's standard heavy bomber for a couple of years after the war until new types were developed, while numerous new variants were created by conversion. They and their roles are described in the previous chapter.

One of the most important of the B-29's postwar roles was the continuation of atomic testing, performed by the 509th Composite Group, the only SAC unit capable of performing such activities in the immediate postwar period. Airborne tests were conducted at Bikini Atoll in the South Pacific. Superfortress 44-27354 *Dave's Dream* became the third of its type to drop an atomic bomb, a 'Fat Man' plutonium device on 1 July 1946. As *The Great Artiste*, this B-29 had previously flown on both of the atomic raids on Japan, carrying instrumentation which recorded data for analysis.

Once More To Battle

Despite the gradual winding down of front line B-29 activities in the late 1940s and early 1950s, the Boeing bomber still had one more period of combat to face, this time in the Korean War, a conflict which had the effect of extending the now obsolescent B-29's active service life longer than had been planned.

On 25 June 1950 communist North

'Enola Gay' photographed after the war carrying the markings of the 313th BW/6th BG despite actually being with the 509th Composite Group.

Korean forces crossed into democratic South Korea, prompting the infant United Nations to quickly respond to meet the threat in what was its first major test. The USAF immediately committed considerable air power to the UN effort including the B-29s of SAC's 22nd and 92nd Bomb Wings plus the 19th Bomb Group of the Far Eastern Air Force. Other USAF B-29 units which would operate in Korea were the 98th and 307th Bomb Wings and the 31st (later 91st) Strategic Reconnaissance Squadron flying RB-29s. The Superfortresses were based at Kadena and Yokata in Japan.

The first B-29 raid was recorded on 13 July when 50 aircraft attacked the port and railway facilities at Wonsan, thus beginning a programme of both strategic and tactical bombing designed to eliminate the North's ability to successfully wage war and to support its troops on the ground. The 31st SRS's RB-29s operated alongside the bombers, gathering photographs and radarscope 'images' which were used for planning and briefing.

The war went very well for the first four months with ground which had been lost in South Korea regained and an invasion of North Korea underway with the capital, Pyongyang, in allied hands and a push towards the Yalu River (the Chinese border) in hand by October.

In the air, the might of the UN forces – led by the substantial American presence – was virtually unchallenged with bombers and strike aircraft able to attack any target with virtual impunity. The North Korean Air Force (NKAF) possessed only modest strength with a fighter regiment comprising about 70 Yak-9 and La-7/11 piston engined fighters, 63 Il-2 ground attack aircraft and a few Yak-18 and Po-2 trainers.

Opposing them were US F-82 Twin Mustangs night fighters, F-51 Mustang and F-80 Shooting Star fighter-bombers, B-26 Invader attack aircraft and the B-29s. Allied air forces were also involved, Australia and South Africa contributing Mustangs (Meteors and Sabres, respectively, later on) while the Republic of Korea Air Force (ROKAF) itself had Mustangs. To this must be added substantial naval air power from the USA initially and then Britain and Australia. The NKAF was no match for this and was quickly nullified.

This left the United Nations air forces with complete air superiority and freedom of movement. The B-29s were particularly effective, ranging freely over North Korean strategic targets and destroying no fewer than 18 of them within two

B-29s of the 92nd BG on their way to a target during the Korean War. The first B-29 raid of the conflict was recorded on 13 July 1950 when the port and railway facilities at Wonsan were attacked.

USAF B-29s enjoyed a free reign over Korea until the Soviet built MiG-15 fighter joined the fray.

months of the start of hostilities. Steel plants, railway yards, harbour facilities, explosives factories and chemicals plants all came under intense bombardment from the Superfortresses, which operated almost with a sense of arrogance as they attacked targets at will and without challenge.

A Rude Awakening

This cosy situation changed dramatically in late October 1950 when allied troops were suddenly confronted by large numbers of Chinese soldiers. The entry of China into the war came as a complete surprise and immediately put the UN forces into retreat. From here, the battle would be long and hard and was not officially declared over until July 1953. At the end of it all, Korea remained divided.

China's intervention brought with it a nasty surprise in the air with the first appearance in November 1950 of the hitherto unknown Soviet built

B-29 in trouble, photographed by the camera gun of a MiG-15 over Korea. The surprise appearance of the MiG-15 forced the USAF to send F-86 Sabres to Korea to help defend the bombers.

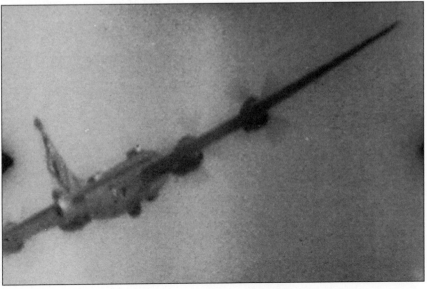

An RAF Washington B.1 whilst on detachment to the Royal Australian Air Force for use at the Woomera Rocket Range in South Australia.

MiG-15 swept wing jet fighter. Also unknown at the time was the fact that these Chinese MiGs were often being flown by Soviet pilots.

Suddenly there was a major threat to the B-29s as their security could no longer be guaranteed as even the F-80 Shooting Star – very much a first generation fighter – was no match for the fast, nimble and well armed MiG-15. B-29 losses mounted, but a solution was found in rushing the North American F-86 Sabre jet fighter to Korea. This much more modern design (regarded by most as the best of its era) was able to take on the MiGs as they rose to intercept the B-29s which from 1951 began to concentrate more on attacking enemy airfields.

Sabres or no Sabres, the B-29 force was never entirely secure for the remainder of the Korean War despite the use of electronic countermeasures, the cover of darkness, radar bombing, black camouflage, varying attack heights, irregular scheduling of raids and different formations. The MiGs were not the only problem, as enemy radar anti aircraft gunlaying and searchlight techniques

improved, despite the electronic countermeasures used against them. There was also the fact that by the early 1950s the Superfortress was very much a World War II aircraft operating in a much changed and more modern environment.

An interesting new weapon was tried out on some B-29s in Korea. Called 'Razon', it was a radio controlled 1,000lb (454kg) bomb which was dropped from a B-29 and then guided to its target by the aircraft's bombardier. The name was derived from the fact that the controller could manipulate Range and AZimuth ONly once it had left the aircraft. Razon proved to be moderately successful, aircraft from the 19th BG destroying 15 bridges with them in 1950-51.

Razon's successor was a 2,000lb (907kg) weapon called 'Tarzon'. Unwieldy and filled with unstable RDX explosive left over from World War II, it was very dangerous to the host B-29 and two were lost trying to ditch the bombs. Thirty Tarzons were dropped in Korea, destroying only six bridges and the weapon was withdrawn in April 1951.

The last B-29 operational mission

in Korea was performed on 27 July 1953, the day the ceasefire came into effect. This was a leaflet dropping sortie flown by RB-29s of the 91st SRS.

Korea saw the end of the Superfortress's combat career. During three years of operations 21,328 combat sorties were flown and 167,000 tonnes of ordnance dropped on a wide variety of targets ranging from factories and fuel dumps to airfields and troop emplacements. Of the sorties flown, just under 2,000 were purely reconnaissance. The cost was 34 B-29s lost, 16 to enemy fighters, four to anti aircraft fire and the remainder to operational accidents.

RAF WASHINGTONS

Only one other air arm operated the B-29, Britain's Royal Air Force. Known as the Washington B.1 in British service, the aircraft was flown in RAF colours as a stopgap pending deliveries of the English Electric Canberra jet bomber.

The Washingtons were delivered in 1950 on 'loan' from the US government. They were serialled WF434-448, WF490-514, WF545-574, WW342-356

The end of the line. Mothballed B-29s await there fate, most likely the hands of the scrapper. (Boeing)

Gathering of the clan. A postwar gathering of surviving B-29s at Tinker AFB, Oklahoma. Some 90 Superfortresses are visible in this photograph. (Boeing)

and WZ966-968 and served with nine squadrons, Nos 15, 35, 44, 57, 90, 115, 149, 192 and 207 before the survivors were returned to the USA in 1954.

No 192 Squadron's activities were particularly interesting as this unit formed part of No 90 (Signals) Group which indulged in electronic warfare and intelligence gathering activities. The squadron is known to have conducted missions around (and possibly over?) the Soviet Union and its

Eastern European allies, these being logged in the Operations Records Books as "Air Ministry Operations".

The USAF also sent B-29s and B-50s over Eastern Europe and Russia on similar missions although they were unfortunate enough to have six Superfortresses shot down between 1952 and 1956 near Soviet territory to the north of Japan and over Manchuria.

Two of the RAF's Washingtons (WW353 and WW354) saw brief serv-

ice with the Royal Australian Air Force from 1952 when they were used for weapons and other trials at the Woomera Rocket Range in South Australia on behalf of the British Ministry of Supply. Although allocated RAAF serial prefixes (A76) both aircraft retained their RAF serials whilst in Australia. They were withdrawn from service in 1956 after flying limited hours and after the removal of useful equipment were sold for scrap in Australia the following year.

The B-29 was a very large aircraft for its day, but it is positively dwarfed by the Convair B-36 strategic bomber of the early Cold War era. At 230 feet (70m), the B-36's wing span was nearly two-thirds more than the B-29's and its length was a similar proportion greater. Maximum weight was nearly three times that of the B-29.